MW00358382

MODERN HEURISTIC TECHNIQUES FOR COMBINATORIAL PROBLEMS

MODERN HEURISTIC TECHNIQUES FOR COMBINATORIAL PROBLEMS

Edited by

COLIN R REEVES BSc, MPhil

Department of Statistics and Operational Research
School of Mathematical and Information Sciences
Coventry University

Halsted Press: an Imprint of
JOHN WILEY & SONS, INC.
NEW YORK TORONTO

Published in the Americas by Halsted Press,
an Imprint of John Wiley & Sons, Inc.,
605 Third Avenue, New York, NY 10158-0012.

Library of Congress Cataloging-in-Publication Data

Modern heuristic techniques for combinatorial problems/Colin R.
Reeves [editor].
p. cm.
Includes bibliographical references and index.
ISBN 0-470-22079-1
1. Programming (Mathematics) 2. Combinatorial optimization.
I. Reeves, Colin R.
QA402.5.M62 1993
519.7—dc20 92-33335
CIP

First published in Great Britain in 1993 by
Blackwell Scientific Publications, Osney Mead, Oxford OX2 0EL

Printed in Great Britain.

Contents

Preface

This book grew out of a conversation at the 32nd Operational Research Society conference at Bangor, North Wales, in 1990. We realised that despite the interest in heuristic techniques, as evident at that conference and in the major OR journals, there was no book-length treatment of the subject. In particular, there seemed to be a need for a comprehensive review of some of the most recently developed strategies. There were some excellent recent works on the subject of Simulated Annealing and Genetic Algorithms, but the developments in Tabu Search, and in the application of Artificial Neural Networks to optimization, had not (and, by and large, still have not) appeared in book form. All these techniques have been inspired in some sense by the realization that attempting to imitate natural processes can bring valuable insights to the problem of combinatorial optimization.

The exploration of the methods stimulated in this way is still at a fairly early stage, whereas the techniques associated with Lagrangean Relaxation have been known and investigated for more than two decades. Yet there is no book devoted exclusively to the subject, and its use as a heuristic tool has not always been fully emphasized. Further, Lagrangean Relaxation can also perform a valuable function in helping to evaluate the quality of solution obtained by other heuristic methods. We therefore thought it would be useful to provide a summary of work in this rather more traditional area alongside the more recent techniques.

I have been fortunate in finding a team of experienced researchers to provide a comprehensive introduction to these areas. My job as overall editor and co-ordinator has been made relatively easy by the quality of their contributions. My only real problem has been to get some of them to stop writing! We have tried to adhere to some common principles and structure in our presentations, but there are inevitably some stylistic differences. It is the intention that each chapter could be read on its own, which has meant a small degree of duplication. Also, where there are obvious points of contact between different techniques, these have been highlighted as they occur.

For the same reason, we have given each chapter its own bibliography, rather than collecting all references into a single list. In the bibliographies, where abbreviations have been used for journal titles, we have followed the list of abbreviations printed in every issue of the *International Abstracts of OR*. Those journals not on this list have had their titles referenced more fully.

It has been our endeavour to make this material as accessible as possible, and so complicated theoretical development has been kept to a minimum. There are areas where some theory is vital, such as in the development of methods based on Artificial Neural Networks, but we would hope that much of this book could be understood at advanced undergraduate level, although we expect it to be of most relevance to post-graduates.

Acknowledgements

Clearly, many others have contributed in no small way to this book. I would like to thank my head of department, Roger Braithwaite, and my other colleagues at Coventry, for arranging some relaxation from my other duties while I was engaged in this task. Discussions with Vic Rayward-Smith at East Anglia have been very helpful, and I am greatly indebted to him and his colleague Geoff McKeown for reading an earlier draft. My colleague Nigel Steele also performed this task, and I am grateful for his comments, as well as for the many stimulating conversations we have had in the course of our own research in Genetic Algorithms. I would also like to thank Colin Paterson and Peter Barnes for their help in overcoming software problems involving LaTeX and PostScript respectively.

Two ladies deserve a mention: Helen Karatza of the Aristotle University of Thessaloniki who gave me a wealth of enlightening etymological detail on the word 'heuristic' and others, and Julia Burden at Blackwell Scientific Publications, who has always answered promptly and helpfully what were probably rather naive queries about preparing a manuscript.

Finally, of course, I am very appreciative of the patience and support of my wife, and the forbearance of our two young children during the preparation of this work.

Colin Reeves
Rugby, November 1992

The Authors

John Beasley studied Mathematics at Cambridge University before going to Imperial College, London for first an MSc, and then a PhD, in Management Science. He has been a lecturer at Imperial College since 1978. His primary research interests are in combinatorial optimization and in its application to industrial problems, especially within the areas of manufacturing and logistics, and he is the author (or co-author) of over 30 articles in refereed journals.

Kathryn Dowsland has a BSc in Pure Mathematics, and the degrees of MSc and PhD in Operational Research from the University of Wales. She worked in the steel industry as an OR Officer for 3 years before joining the University of Wales in Swansea, where she has been a Lecturer in the European Business Management School since 1985. Her research interests include graph-theoretic approaches to problem-solving and the use of tabu search and simulated annealing, particularly as they relate to packing, scheduling and timetabling.

Fred Glover is US West Chair of Systems Science at the University of Colorado at Boulder, where he has developed the concept of tabu search. His research interests are wide and he is the author of over 200 papers in the areas of combinatorial optimization, computer science and artificial intelligence. He has been featured as a National Visiting Lecturer by the Institute of Management Science and the Operations Research Society of America, and has served as a host and lecturer in the US National Academy of Sciences Program of Scientific Exchange. He has held editorial posts on several journals in the US and abroad, has received numerous awards and honorary fellowships, and serves on the advisory board of several organizations.

Manuel Laguna is an Assistant Professor in Management Science at the University of Colorado, having received master's and doctor's degrees in Operations Research and Industrial Engineering from the University of Texas at Austin. His research centres on the interface between OR and AI and resulting applications in the areas of production scheduling, telecommunications and facility layout, on which he has published several papers.

Carsten Peterson received his PhD in theoretical physics in 1976 from the University of Lund, Sweden where he has been Associate Professor in the Dept of Theoretical Physics since 1983, and he has also held posts in Copenhagen and Stanford. In 1991 he was awarded the Goran Gustafsson Prize in Physics. His original research interests were in elementary particle physics, but since 1987 he has concentrated on optimization methods based on the exploitation of artificial neural networks, where he has made many important and original contributions.

Colin Reeves obtained his BSc in Mathematics before carrying out research on vehicle routing problems for a MPhil degree at the then Lanchester Polytechnic in 1976. After working for the Ministry of Defence as an OR scientist where he developed an interest in pattern recognition problems, he returned to what is now Coventry University, joining the Department of Statistics and OR in 1978. He is currently a Senior Lecturer in OR. His main research interests are in pattern recognition, in combinatorial optimization in general, and in heuristic methods in particular, on which he has published several papers. His most recent publications are mainly in the area of Genetic Algorithms, especially in their applications to Neural Networks.

Bo Söderberg received his PhD in theoretical physics in 1984 from the University of Lund. He has held positions in Copenhagen, at CNRS, Marseille, and in Bielefeld, Germany. He returned to Lund in 1988, where he is currently an Associate Professor in the Department of Theoretical Physics. After doing research in elementary particle physics and non-linear dynamics, he too became interested in artificial neural networks—in particular, in developing the mean field algorithm for optimization, although he has also made several other contributions to this swiftly-developing subject.

Chapter 1

Introduction

Colin R Reeves and John E Beasley

1.1 Combinatorial Problems

In the past 40 years many researchers have studied the problem of finding optimal solutions to problems which can be structured as a function of some *decision variables*, perhaps in the presence of some *constraints*. The subject is very wide, and many books have already been written on its various aspects. Such problems can be formulated generally as follows:

$$\begin{aligned} \textit{Minimize} \quad & f(\mathbf{x}) \\ \textit{subject to} \quad & g_i(\mathbf{x}) \geq b_i; \quad i = 1, \ldots, m; \\ & h_j(\mathbf{x}) = c_j; \quad j = 1, \ldots, n. \end{aligned}$$

Here, $\mathbf{x}$ is a vector of decision variables, and $f(\cdot)$, $g_i(\cdot)$ and $h_j(\cdot)$ are general functions. (Actually the equality constraints are not strictly necessary, as they can be formulated in terms of pairs of inequalities, but it is often helpful to make them explicit.) This formulation has assumed the problem is one of minimization, but the modifications necessary for a maximization problem are clear.

There are many specific classes of such problems, obtained by placing restrictions on the type of functions under consideration, and on the values that the decision variables can take. Perhaps the most well-known of these classes is that obtained by restricting $f(\cdot)$, $g_i(\cdot)$ and $h_j(\cdot)$ to be linear functions of decision variables which are allowed to take fractional (continuous) variables, which leads to problems of *linear programming.*

In this book, we consider another class of problems: those of a *combinatorial* nature. This term is usually reserved for problems in which the decision variables are discrete—i.e. where the solution is a set, or a sequence, of integers or other discrete objects. The problem of finding optimal solutions to such problems is therefore known as *combinatorial optimization.*

Some examples of this kind of problem are as follows:

Example 1.1 (The assignment problem) *A set of n people is available to carry out n tasks. If person i does task j, it costs c_{ij} units. Find an assignment $\{\pi_1, \ldots, \pi_n\}$ which minimizes*

$$\sum_{i=1}^{n} c_{i\pi_i}.$$

Here the solution is represented by the permutation $\{\pi_1, \ldots, \pi_n\}$ of the numbers $\{1, \ldots, n\}$.

Example 1.2 (The 0-1 knapsack problem) *A set of n items is available to be packed into a knapsack with capacity C units. Item i has value v_i and uses up c_i units of capacity. Determine the subset I of items which should be packed in order to maximize*

$$\sum_{i \in I} v_i$$

such that

$$\sum_{i \in I} c_i \leq C.$$

Here the solution is represented by the subset $I \subseteq \{1, \ldots, n\}$.

Example 1.3 (The set covering problem) *A family of m subsets collectively contains n items such that subset S_i contains $n_i (\leq n)$ items. Select $k (< m)$ subsets $\{S_{i_1}, \ldots, S_{i_k}\}$ such that $|\cup_{j=1}^{k} S_{i_j}| = n$ so as to minimize*

$$\sum_{j=1}^{k} c_{i_j}$$

where c_i is the cost of selecting subset S_i.
Here the solution is represented by the family of subsets $\{S_{i_1}, \ldots, S_{i_k}\}$.

Example 1.4 (The vehicle routing problem) *A depot has m vehicles available to make deliveries to n customers. The capacity of vehicle k is C_k units, while customer i requires c_i units. The distance between customers i and j is $d_{i,j}$. No vehicle may travel more than D units. Allocate customers to vehicles and find the order in which each vehicle visits its customers so as to minimize*

$$\sum_{k=1}^{m} \sum_{i=0}^{n_k} d_{\pi_{i,k}, \pi_{i+1,k}}$$

such that

$$\sum_{i=1}^{n_k} c_{\pi_{i,k}} \leq C_k; \; k = 1, \ldots, m$$

$$\sum_{i=0}^{n_k} d_{\pi_{i,k}, \pi_{i+1,k}} \leq D; \; k = 1, \ldots, m$$

$$\sum_{k=1}^{m} n_k = n$$

Here, vehicle k visits n_k customers, and the solution is represented by the permutation $\{\pi_{1,1}, \ldots, \pi_{n_1,1}, \ldots, \pi_{1,m}, \ldots, \pi_{n_m,m}\}$ of the numbers $\{1, \ldots, n\}$, which is partitioned by the numbers $\{n_k\}$. It is also understood that the depot is represented by the 'customers' $\pi_{0,k}$ and $\pi_{n_k+1,k}$ for each k.

It should be clear that in these examples, we are using the term 'problem' in a generic sense; in a given real-life situation, we will have a *particular* problem, in which the symbols used to describe it take specific numerical values. It is customary to use the term *instance* to distinguish between the particular and the general situation.

1.1.1 Links with linear programming

Combinatorial problems, such as those described above, have close links with linear programming (LP), and most of the early attempts to solve them used developments of LP methods, generally by introducing integer variables taking the values 0 or 1, in order to produce an integer programming (IP) formulation. For example, in the case of the 0-1 knapsack problem, we define

$$\begin{aligned} x_i &= 1 \text{ if item } i \text{ is packed} \\ &= 0 \text{ otherwise.} \end{aligned}$$

The problem then reduces to the following Integer Program:

$$maximize \sum_{i=1}^{n} x_i v_i$$

$$such\ that \sum_{i=1}^{n} x_i c_i \leq C.$$

This is a particularly straightforward formulation, as is the set covering problem, whose IP formulation will be used extensively in chapter 6. However, in many cases it needs some ingenuity to find an IP formulation of a combinatorial optimization problem. (Any reader who doubts this might like to try formulating the vehicle routing problem outlined above as an IP!) Moreover, such formulations often involve very large numbers of variables and constraints, and general-purpose IP computer codes cannot usually cope with very large problems. IP is actually much 'harder' than ordinary LP, for reasons which we shall discuss later.

However, although IP is not in general a successful route to finding optimal solutions to combinatorial problems, there are good reasons for its popularity. Firstly, the act of formulation is itself often helpful in defining more precisely the nature of a given problem. Secondly, for many problems, it is possible within IP to use the technique of Lagrangean relaxation to generate a lower bound on the optimal solution. Such information can be very valuable, as we shall discover.

1.2 Local and Global Optima

A feature of many combinatorial problems is that while there may be several true or 'global' optima, there are many more that are in some sense only 'locally' optimal. We can make this idea more concrete by introducing the concept of a *neighbourhood*.

Roughly speaking, a neighbourhood $\mathbf{N}(\mathbf{x}, \sigma)$ of a solution $\mathbf{x}$ is a set of solutions that can be reached from $\mathbf{x}$ by a simple operation σ. Such an operation σ might be the removal of an object from, or addition of an object to, a solution. The interchange of two objects in a solution is another example of such an operation which is particularly common in sequencing problems. Often, and particularly in Tabu Search, these operations are called *moves.* If a solution $\mathbf{y}$ is better than any other solution in its neighbourhood $\mathbf{N}(\mathbf{y}, \sigma)$, then $\mathbf{y}$ is a local optimum with respect to this neighbourhood.

In some cases it is possible to find a move σ such that a local optimum is also a global optimum. For example, in some one-machine sequencing problems, a *pairwise interchange* move has this property. However, for many problems this is not possible, and it is necessary to use some form of *implicit enumeration.*

Of course, there are many existing treatments of combinatorial optimization —[1, 2, 3, 4] for instance—but these tend to concentrate on methods which are *exact* rather than heuristic. That is, they are mainly concerned with those techniques which *guarantee* to find the optimal solution to a stated problem. These methods usually rely on links with the theories of linear programming or graphs, or else use an implicit enumeration approach such as branch-and-bound or dynamic programming. These are not the concern of this book.

1.3 Heuristics

The techniques described in this work are ones which have become known by the term *heuristic.* This term derives from the Greek *heuriskein* meaning to find or discover. When Archimedes jumped out of his bath he used the past tense of this verb: *eureka* means 'I've found (it)'. We could hardly claim Archimedes as the father of heuristics, however, and in fact, the etymological perspective is not very helpful, as White [5] for example has pointed out. The word has indeed been used in Artificial Intelligence circles with quite a different connotation—Pearl [6], for example, uses the term to include methods such as branch-and-bound which find globally optimal solutions, and this usage probably pre-dates the now common one. In fact, this usage is perfectly reasonable; if a heuristic is a 'finding' method, we are entitled to ask *what* it is finding, if not a global optimum. What in Operational Research is now almost universally called a 'heuristic' would be better described as a 'seeking' method, as it cannot guarantee to *find* anything[1].

Nevertheless, in the usage that has become common in the context of combinatorial optimization, the term *heuristic* is used as a contrast to methods which guarantee to find a global optimum. The first recorded usage of the term in the *International Abstracts of OR* seems to have been in 1960 (in a paper written in French), but by the mid-1960s, the currently accepted connotation seems to have become well-established. In this book, we will stay with this usage, and use the following working definition:

Definition 1 *A* heuristic *is a technique which seeks good (i.e. near-optimal) solutions at a reasonable computational cost without being able to guarantee either feasibility or optimality, or even in many cases to state how close to optimality a particular feasible solution is.*

A large amount of work has been carried out on heuristic methods for solving combinatorial problems; a recent survey by Zanakis *et al.* [7] listed over 400 papers in which a heuristic was used. Anyone who has followed the development of heuristics over the past two decades would have no difficulty in adding many papers that were omitted from this survey. Furthermore, there have been several hundred papers published since the survey was carried out.

The causes of this explosion of interest would seem to be two-fold. On the one hand, the development of the concept of *computational complexity* has provided a rational basis for exploring heuristic techniques rather than pursuing the Holy Grail of optimality, while at the same time there has been a significant increase in the power and efficiency of the more modern approaches. Twenty years ago an eminent person in Operational Research circles suggested to the editor of this book that using a heuristic was 'an admission of defeat'; times have certainly changed for heuristics are now considered very respectable. The reason for this change in attitude is discussed below.

---

[1] If we wanted to stay with the Greek, we could have used a term like *zetetic*, meaning 'seeking', and allowed the earlier use of the word heuristic to stand. However, it is about 30 years too late for inventing a more consistent vocabulary!

1.3.1 The case for heuristics

A naive approach to solving an instance of a combinatorial problem is simply to list all the feasible solutions of a given problem, evaluate their objective functions, and pick the best. However, it is immediately obvious that this approach of *complete enumeration* is likely to be grossly inefficient; further, although it is possible *in principle* to solve any problem in this way, *in practice* it is not, because of the vast number of possible solutions to any problem of a reasonable size. To illustrate this point, consider the famous travelling salesman problem (TSP).

This problem has exercised a particular fascination to researchers in combinatorial optimization, probably because it is so easily stated, yet so hard to solve. For the record, the problem is as follows: a salesman has to find a route which visits each of N cities once and only once, and which minimizes the total distance travelled. As the starting point is arbitrary, there are clearly $(N-1)!$ possible solutions (or $(N-1)!/2$ if the distance between every pair of cities is the same regardless of the direction of travel). Suppose we have a computer that can list all possible solutions of a 20 city problem in 1 hour. Then, using the above formula, it would clearly take 20 hours to solve a 21-city problem, and 17.5 days to solve a 22-city problem; a 25-city problem would take nearly 6 centuries. Because of this exponential growth in computing time with the size of the problem, complete enumeration is clearly a non-starter.

In the early days of Operational Research, the emphasis was mostly on finding the optimal solution to a problem—or rather, to a *model* of a problem which occurred in the real world. To this end, various exact algorithms were devised which would find the optimal solution to a problem much more efficiently than complete enumeration. The most famous example is the *Simplex* algorithm for linear programming problems[2]. At first, while such algorithms were capable of solving small instances of a problem, they were not able to find optimal solutions to larger instances in a reasonable amount of computing time. As computing power increased, it became possible to solve larger problems, and researchers became interested in how the solution times varied with the size of a problem. In some cases, such

[2]Actually, Simplex is now known to show exponential behaviour in the worst-case, although on average its performance is excellent.

as the 'Hungarian' method [8] for solving the assignment problem, or Johnson's method [9] for 2-machine sequencing, the computing effort could be shown to grow as a low-order polynomial in the size of the problem. However, for many others, such as the travelling salesman problem, the computational effort required was an exponential function of the problem size.

In a sense, therefore, these exact methods (such as branch-and-bound or dynamic programming) perform no better than complete enumeration. The question that began to exercise the minds of researchers in the late 1960s was the following: is there a 'polynomial' optimising algorithm for a problem such as the TSP? Nobody has yet been able to answer this question one way or the other, but in 1972, Karp [10] was able to show that if the answer to this question is 'yes' for the TSP, then there is also a polynomial algorithm for a number of other 'difficult' problems. As no such algorithm has yet been found for any of these problems, it strongly *suggests* that the answer to the original question is 'no'. However, the actual answer is still unknown, some 20 years after Karp's original work.

P and NP

Technically, those problems such as assignment which have a known polynomial algorithm are said to be in class **P**. What of the TSP and other 'difficult' problems: are they inherently of exponential complexity? It turns out that not many problems can be *proven* to be this bad; rather most of these 'difficult' problems are in class **NP**—an abbreviation for 'non-deterministic polynomial'. In fact, the TSP is one of a large class of problems which are, in a sense, the most difficult problems of all problems in **NP**—this sense being that if a polynomial algorithm were to be found for any of them, it would mean that a polynomial algorithm existed for all problems in **NP**.

Actually things are a little more complicated than this rough description implies. The concept of a *non-deterministic polynomial* algorithm is a fairly subtle one, but it is possible to view it intuitively as follows. First we observe that problems in **NP** are really 'decision' or 'recognition' problems: that is, rather than ask for the optimal-length tour of a TSP, say, we ask *'is there a tour of length less than L'*? The 'recognition' and 'optimization' versions of a problem are closely related, since it is intuitively obvious that an algorithm for the recognition version of a problem can be employed to solve the optimization version. For example we could specify a range $[0, U]$ in which the optimal value is known to lie; we would then ask if there is a solution in $[0, U/2]$, and by repeatedly halving the range in which the optimum lies we would eventually find the optimal value. However, we would not usually try to solve an instance of an optimization problem by this means!

We now imagine a computer which has the property that each time it faces a choice it divides into two 'copies' of itself, with each copy being explored in parallel. It is thus rather like a tree-search where all branches can be searched simultaneously. If (and only if) one of the copies answers the recognition problem in the affirmative, will we have solved the recognition problem. If the maximum time taken by a branch is polynomially bounded, then the problem is in **NP**. Another way of looking at this intuitively is to suppose that we could 'guess' a solution to the problem, and require that checking the answer could be carried out in polynomial time.

Next, we need the concept of *transformability* (sometimes called *reducibility*). Suppose we have a problem P_1 which can be solved by an algorithm A. If we can transform every instance of another problem P_2 into an instance of P_1 in *polynomial* time, then clearly we can use this fact, plus the algorithm A, to solve P_2. The class of instances of P_1 is at least as wide as (and probably wider than) the class of transformed instances of P_2—the latter are in a sense 'special cases' among the instances of P_1. Thus it is reasonable to regard P_1 as being at least as hard as (and possibly harder than) P_2.

If a problem P is such that *every* problem in **NP** is polynomially transformable to P, we say that P is *NP-hard.* If in addition problem P itself belongs to **NP**, P is said to be *NP-complete.* Bearing in mind what we have said above, the implication is that such problems are the 'hardest' of all problems in **NP**. NP-complete problems are clearly important, since if we were to find a polynomial algorithm for any one of them, we would have found a polynomial algorithm for all problems in **NP**; that is we would have shown that **P=NP**. The class of NP-complete problems is now known to be a large one; a list of problems proven to be such is contained in [12]—a list which has been periodically updated in the *Journal of Algorithms.*

Readers of the literature on combinatorial optimization will observe that there is sometimes a confusion of terminology. For example, *optimization* problems as such are not in **NP**, even when their decision versions are. (The 'binary search' procedure sketched above, although valid in itself, cannot show that an optimization problem is in **NP**, since we would require it at least some of the time to return a *negative* answer to the recognition problem, which may not take polynomial time). Despite this, many people (including some of the contributors to this book!) refer to a problem as being NP-complete, even when it is the optimization version which is under consideration. Although precision in terminology is important, by a statement such as 'this (optimization) problem is NP-complete', an author is probably assuming, reasonably enough, that the reader will understand him to mean 'the decision version of this optimization problem is NP-complete'. Nevertheless, many authors make use of the term NP-hard for describing optimization problems, and that practice is to be preferred.

This account has been deliberately intuitive and informal, following the approach of introductory texts such as that by Manber [11]. Readers who would like a more formal and rigorous description are recommended to study the books by Papadimitriou and Steiglitz [2] or Garey and Johnson [12], which provide a detailed account of this fascinating area of computational complexity. Some of the recent developments with particular reference to optimization problems are reviewed by Bruschi *et al.* [13]. Other useful references are Karp [14] and Rayward-Smith [15], while at a more popular level the articles by Kolata [16] and Lewis and Papadimitriou [17] will also help to make things clearer.

For the purposes of this book, a formal approach would not be appropriate, since we wish to focus attention on the *practical effect* of the theoretical results concerning computational complexity. All attempts to prove that **P**=**NP** theoretically have so far failed. Further, because no exact polynomial algorithm has been found for any problem in **NP**, despite many person-centuries of effort by many very distinguished and very clever people, there is strong circumstantial evidence that **P**≠**NP**, which argues in favour of finding alternative ways of 'solving' difficult problems. There is always the possibility that someone will prove that **P**=**NP**, but the chance would seem to be a very small one. Until someone does find such a proof, the use of heuristics has considerable justification.

Models and problems

There is another argument in favour of using heuristics: what we are actually optimising is a *model* of a real-world problem. There is no guarantee that the best solution to the model is also the best solution to the underlying real-world problem. To put it another way, should we prefer an exact solution of an approximate model, or an approximate solution of an exact model? Of course, we cannot hope for a truly exact model, but heuristics are usually rather more flexible and are capable of coping with more complicated (and more realistic) objective functions and/or constraints than exact algorithms.

This is the case for methods like simulated annealing, tabu search and genetic algorithms, for example, where objective functions need no simplifying assumptions of linearity. Thus it may be possible to model the real-world problem rather more accurately than is possible if an exact algorithm is used.

1.3.2 Modern methods

We now outline the contents of this book. Five topics are covered, and by and large they can all be described as 'modern'. In each case, the purpose is to give the reader a general introduction to the basic concepts underlying the particular topic, together with an overview of developments and extensions that have been found useful in improving its effectiveness and efficiency. Occasionally, proofs of important theoretical results will be given in order to illuminate a particular aspect of the method being discussed, but in most cases such results will be quoted without proof as the primary intention of this book is to give practical information. Naturally, full references will be given for those who wish to pursue theoretical studies.

A survey of some of the most important reported applications will also be included to enable readers to decide on an appropriate technique for their own problems. All the techniques rely heavily on computing skills for their practical implementation, and it is hoped that sufficient information will be given that a competent programmer will have little difficulty in carrying out such implementation.

The question may well be asked: how do these particular techniques relate to other heuristics that have been developed over the years, and what are their distinguishing features? There have already been several surveys of heuristic methods: see White [5] and Zanakis *et al.* [7] for example, while the journal *Management Science* recently devoted a whole issue [18] to the subject. (Other surveys and review papers which readers might find useful and interesting are those by Silver *et al.* [19], Zanakis and Evans [20], and Foulds [21].) These tend to classify heuristics into several broad categories: greedy construction methods, neighbourhood search (improvement) routines, relaxation techniques, partial enumeration, decomposition and partition approaches and so on. However, the concern of this book is not with these as such, although some of the techniques described in the chapters that follow could be placed into some of these categories.

Many heuristics are problem-specific, so that a method which works for one problem cannot be used to solve a different one. However, there is increasing interest in techniques which are applicable far more generally, and over the last decade, several techniques with this characteristic have been developed. These have proved to be very powerful when applied to a large number of problems, and it is some of these that are the subject of this book.

Two of these techniques are more powerful developments of the popular heuristic of *local neighbourhood search.* A more formal description of this idea is given later, but in essence it is very simple: the process starts with a (sub-optimal) solution to a particular problem and searches a defined neighbourhood of this solution for a better one. Having found one the process re-starts from the new solution. It continues to iterate in this way until no improvement can be found to the current solution. This final solution is unlikely to be the global optimum—although, with respect to its neighbourhood, it is clearly locally optimal.

One of the earliest examples of this approach is Lin's well-known λ-optimal heuristic for the travelling salesman problem [22], where the neighbourhood is defined by deleting λ edges in the current tour and attempting to re-connect the tour in such a way that the total tour length is reduced. Similar approaches have been reported for many other combinatorial problems.

Various methods of dealing with local optimality have been used. An obvious approach is to enlarge the neighbourhood—for example, by moving from a 2-optimal heuristic to a 3-optimal one. Another popular idea is to start the whole procedure again from different initial solutions, in the hope that it will converge to different local optima. The best of the many solutions thus found is then chosen.

A third option which has gained recent attention is to allow 'uphill moves', whereby the solution from which the search is re-started is allowed to be worse than the previous one. There has to be some restriction on accepting such moves, otherwise the procedure would amount to a search of the whole solution space, but by using this idea, there is a chance that the process can 'climb out' of a local optimum and find a better solution.

The methods of simulated annealing and tabu search have received considerable coverage in the last few years, and there have been several applications to various NP-hard problems which have achieved considerable success. These methods are both examples of this third approach to avoiding local optima; they form the subjects of chapters 2 and 3 of this book.

To return to the second part of the question posed earlier: the distinguishing feature of most of the techniques discussed in this book is the way they attempt to simulate some naturally-occurring process. This idea of simulating natural processes has been shown to be of considerable value in solving complex problems; simulated annealing was in fact originally introduced as an analogy to thermodynamic processes, while tabu search finds some of its motivation in attempts to imitate 'intelligent' processes—in particular, by providing heuristic search with a facility that implements a kind of 'memory'. More recently, the concept of a genetic algorithm, the subject of chapter 4, where problems are formulated as an analogy to genetic structures, has been shown to give insight into the solution of optimization problems. The area of artificial neural networks (ANNs) is an attempt to use an analogy with the computational properties of the human brain. Although books on ANNs are proliferating vigorously, much of the work reported here is very recent and has not previously appeared in book form. In chapter 5 some of this research on the way combinatorial problems can be mapped to neural networks will be described.

The final topic of Lagrangean relaxation, considered in chapter 6, has a different origin, and it is also somewhat less 'modern' than the other techniques contained in this book. It is derived from the well-known integer programming paradigm, and when coupled with a branch-and-bound (tree search) algorithm is in principle capable of finding globally optimal solutions. However, in the last decade, Lagrangean relaxation has often been used as a heuristic tool which finds good (but not necessarily optimal) solutions. It also has another significant rôle in providing high-quality bounds on the optimal solution. This means that in cases where a Lagrangean relaxation is possible, the quality of a feasible solution obtained by a heuristic can be assessed, and the need for further computational effort can be judged. For these reasons, we have provided a summary of developments in this rather more 'traditional' area of research on combinatorial optimization.

1.3.3 Evaluation of heuristics

As already remarked, whenever a heuristic is applied, the user always faces the problem of how good the generated solution really is. In some cases it is possible to analyse heuristic procedures explicitly and find theoretical results bearing on their average or worst-case performance. However, analysis of *general* performance in this way is often difficult, and in any case may provide little help in deciding how well a heuristic has performed in a *particular* instance. Information on *specific* performance of this nature is more appropriate to the techniques described in this book. The contents of chapter 6 on Lagrangean relaxation are of course an important element in obtaining such information. The relationship between Lagrangean relaxation and the other methods surveyed in this book is made clearer in the following diagrams.

Suppose that we have some instance of an NP-hard minimization problem. If, as in Figure 1.1, we draw a vertical line representing value (the higher up this line the higher the value) then somewhere on this line is the optimal solution to the problem instance we are considering. Exactly where on this line the optimal solution lies is unknown, but it must be somewhere!

Conceptually therefore this optimal solution value divides our value line into two:

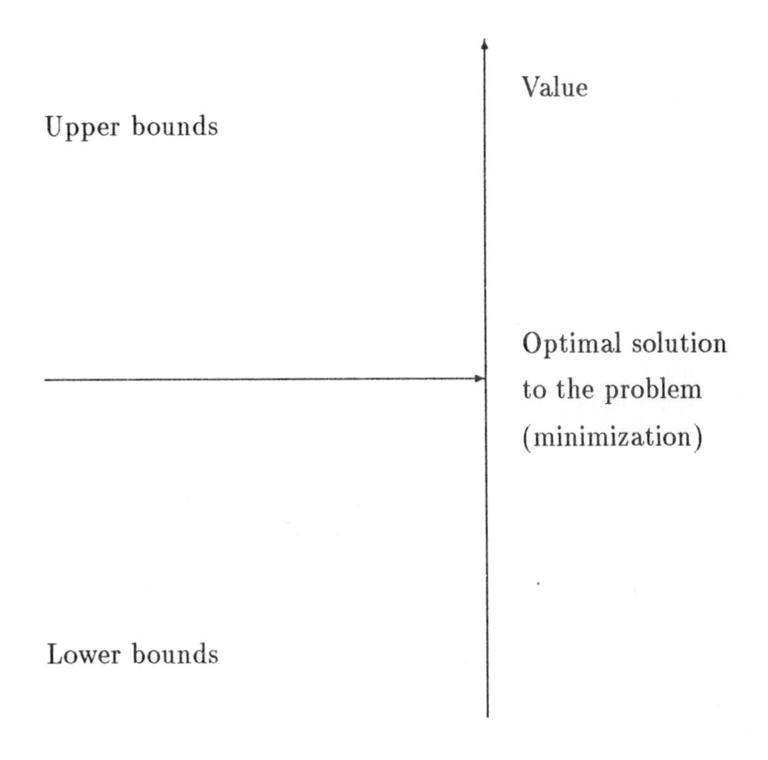

Figure 1.1: Diagrammatic representation of a minimization problem

- above the optimal solution value are upper bounds, values which are above the (unknown) optimal solution value;
- below the optimal solution value are lower bounds, values which are below the (unknown) optimal solution value.

In order to discover the optimal solution value, an algorithm should really address both these issues—it must concern itself with both upper and lower bounds.

In particular the quality of these bounds is important to the computational success of any algorithm:

- we like upper bounds that are as close as possible to the optimal solution, i.e. as small as possible;

- we like lower bounds that are as close as possible to the optimal solution, i.e. as large as possible.

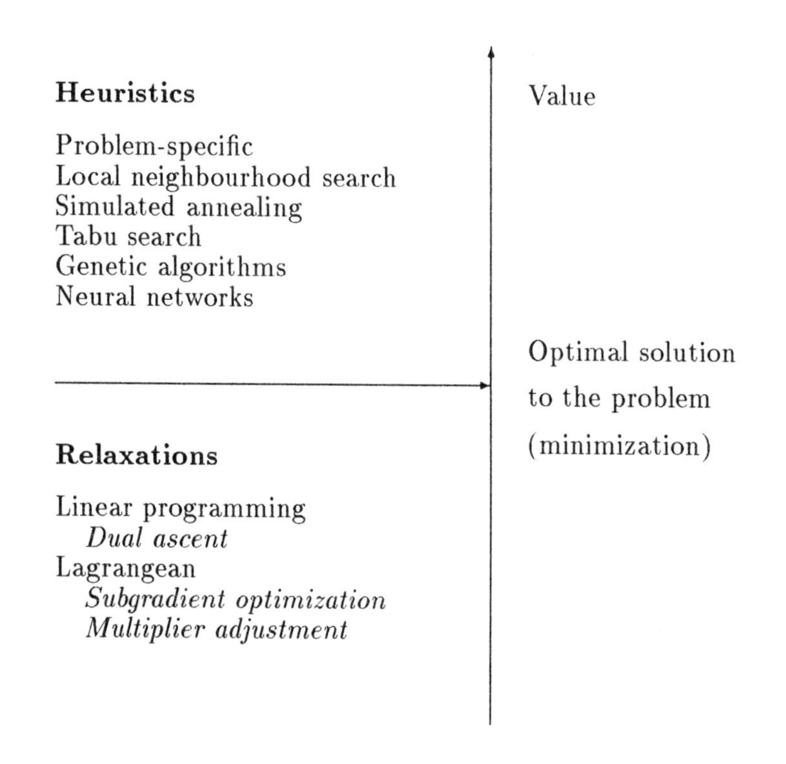

Figure 1.2: Connection between heuristics and relaxation

In the context of a minimization problem, the upper bound is generated by a heuristic method such as one of the techniques described in chapters 2-5. Lower bounds can be generated by linear programming, using the technique of *dual ascent*, for example, or by Lagrangean relaxation, using *subgradient optimization* or *multiplier adjustment*; all of these are discussed in chapter 6. The connection between heuristic methods and relaxation can be summarized by the diagram of Figure 1.2. Note however, that no order relationship is implied in the way the methods are listed in this diagram: for instance, it should not be inferred that genetic algorithms inherently give better bounds than simulated annealing!

However, it is not always easy to find a suitable Lagrangean relaxation, and readers may be interested in swifter ways of assessing a heuristic solution. In chapter 7 some of the approaches that have been found useful in this context are described, together with some brief remarks on worst-case and average performance analysis. These approaches are

- the computation of lower or upper bounds (other than by Lagrangean relaxation);
- empirical performance on test problems; and
- the use of statistical inference.

Thus, the contents of this book should enable the reader not only to find an efficient and effective method of solving a particular problem, but also to *assess* the quality of the solution obtained in an objective manner.

References

[1] N.Christofides, A.Mingozzi, P.Toth and C.Sandi (1979) *Combinatorial Optimization.* John Wiley and Sons, Chichester.

[2] C.D.Papadimitriou and K.Steiglitz (1982) *Combinatorial Optimization: Algorithms and Complexity.* Prentice-Hall, Englewood Cliffs, NJ.

[3] E.Lawler (1981) *Combinatorial Optimization: Networks and Matroids.* Holt, Rinehart and Winston, New York.

[4] E.Lawler, J.K.Lenstra, A.H.G.Rinnooy Kan and D.B.Shmoys (1985) (Eds.) *The Traveling Salesman: A Guided Tour of Combinatorial Optimization.* John Wiley and Sons, Chichester.

[5] D.J.White (1990) Heuristic programming. *IMAJMABI*, **2**, 173-188.

[6] J.Pearl (1984) *Heuristics.* Addison-Wesley, Reading, Mass.

[7] S.H.Zanakis, J.R.Evans and A.A.Vazacopoulos (1989) Heuristic methods and applications: a categorized survey. *EJOR*, **43**, 88-110.

[8] H.W.Kuhn (1955) The Hungarian method for the assignment problem. *NRLQ*, **2**, 83-97.

[9] S.M.Johnson (1954) Optimal two- and three-stage production schedules with setup times included. *NRLQ*, **1**, 61-68.

[10] R.M.Karp (1972) Reducibility among combinatorial problems. *In* R.E.Miller and J.W.Thatcher (Eds.) *Complexity of Computer Computations*, Plenum Press, New York.

[11] U.Manber (1989) *Introduction to Algorithms.* Addison-Wesley, Reading, Mass.

[12] M.R.Garey and D.S.Johnson (1979) *Computers and Intractability: A Guide to the Theory of NP-Completeness.* W.H.Freeman, San Francisco.

[13] D.Bruschi, D.Joseph and P.Young (1989) A structural overview of **NP** optimization problems. *In* H.Djidjev (Ed.) *Optimal Algorithms*, Springer Verlag, Berlin.

[14] R.M.Karp (1986) Combinatorics, complexity and randomness. *Comm. of the ACM*, **29**, 98-117.

[15] V.J.Rayward-Smith (1986) *A First Course in Computability.* Blackwell Scientific Publications, Oxford.

[16] G.B.Kolata (1980) Solve one and you could solve them all. *New Scientist*, 3rd April 1980, 8-10.

[17] H.R.Lewis and C.H.Papadimitriou (1978) The efficiency of algorithms. *Scientific American*, **238(1)**, January 1978, 96-109.

[18] M.L.Fisher and A.H.G.Rinnooy Kan (1988) (Eds.) The design, analysis and implementation of heuristics. *Man.Sci.*, **34**, 263-429.

[19] E.Silver, R.V.Vidal and D.de Werra (1980) A tutorial on heuristic methods. *EJOR*, **5**, 153-162.

[20] S.H.Zanakis and J.R.Evans (1981) Heuristic 'optimization': Why, when and how to use it. *Interfaces*, **11**, 84-91.

[21] L.R.Foulds (1983) The heuristic problem-solving approach. *JORS*, **34**, 927-934.

[22] S.Lin (1965) Computer solutions of the traveling salesman problem. *Bell Systems Tech. J.* , **44**, 2245-2269.

Chapter 2

Simulated Annealing

Kathryn A Dowsland

2.1 Introduction

The use of simulated annealing as a technique for discrete optimization dates back to the early 1980s. It was heralded with much enthusiasm as it appeared to be both simple to implement and widely applicable, and as a result of articles in popular scientific journals [1, 2, 3] researchers from a wide variety of disciplines experimented with it in the solution of their own problems. Thus anyone considering the use of simulated annealing today has access to a wide range of literature covering both theoretical and empirical results. This chapter aims to provide a comprehensive introduction to the technique by bringing together some of this diverse experience. It includes enough basic material to enable a first time user to get started, and then goes on to examine a variety of modifications to the basic method which more experienced users may want to explore. References to some of the more detailed research publications in the field are also provided.

The ideas that form the basis of simulated annealing were first published by Metropolis *et al.* [4] in 1953 in an algorithm to simulate the cooling of material in a heat bath—a process known as annealing. If solid material is heated past its melting point and then cooled back into a solid state, the structural properties of the cooled solid depend on the rate of cooling. For example, large crystals can be grown by very slow cooling, but if fast cooling or quenching is employed the crystal will contain a number of imperfections. The annealing process can be simulated by regarding the material as a system of particles.

Essentially, Metropolis's algorithm simulates the change in energy of the system when subjected to a cooling process, until it converges to a steady 'frozen' state. Thirty years later, Kirkpatrick *et al.* [1] suggested that this type of simulation could be used to search the feasible solutions of an optimization problem, with the objective of converging to an optimal solution.

This approach can be regarded as a variant of the well-known heuristic technique of local (neighbourhood) search, in which a subset of the feasible solutions is explored by repeatedly moving from the current solution to a neighbouring solution. For a minimization problem the traditional forms of local search employ a descent strategy, in which the search always moves in a direction of improvement. However, such a strategy often results in convergence to a local rather than a global optimum. A number of approaches have been suggested which partially overcome this problem. As discussed in chapter 1, it may be possible to repeat the algorithm using several different starting solutions, or to increase the complexity of the neighbourhoods thus widening the scope of the search. Unfortunately, none of these variants have proved to be entirely satisfactory.

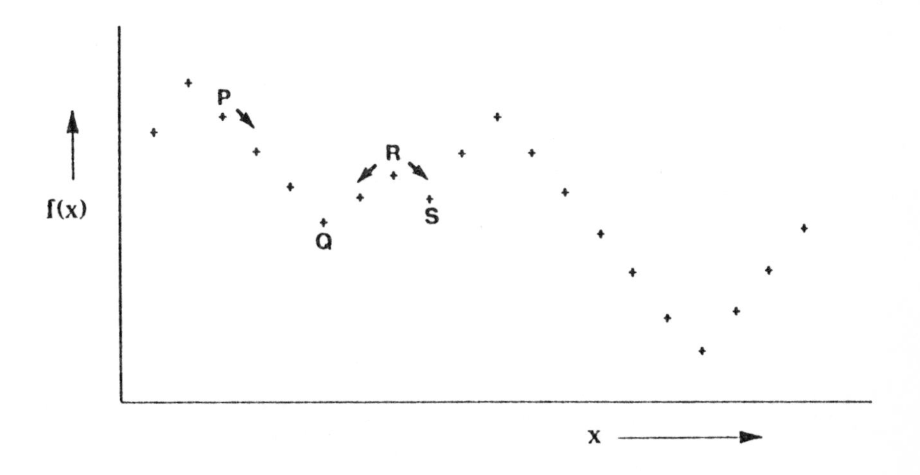

Figure 2.1: Illustration of a descent strategy

The solutions obtained by these descent strategies are totally dependent on the starting solution(s) employed. Figure 2.1 illustrates a simple example of a function in which each solution has just two neighbours, represented by the points immediately to the left and right. A descent strategy will always move towards the bottom of the valley containing the starting point. For example, initial solution P will always lead to terminal solution Q. The only exceptions are the local maxima (e.g. R) where either of 2 valleys (Q and S) may be reached. These observations obviously extend to more complex structures in any number of dimensions.

It is arguable that a reliable heuristic should be less dependent on the starting point. It is clear from Figure 2.1 that this must involve some uphill moves, but as the final objective is to converge to a minimum point these must be used sparingly and in a controlled manner. In the simulated annealing heuristic uphill moves are allowed, but their frequency is governed by a probability function which changes as the algorithm progresses.

The inspiration for this form of control was Metropolis's work in statistical thermodynamics. The laws of thermodynamics state that at temperature t, the probability of an increase in energy of magnitude δE is given by

$$p(\delta E) = \exp(-\delta E/kt) \tag{2.1}$$

where k is a physical constant known as Boltzmann's constant.

Metropolis's simulation generates a perturbation and calculates the resulting energy change. If energy has decreased the system moves to this new state. If energy has *increased*, the new state is accepted according to the probability given in equation 2.1. The process is repeated for a predetermined number of iterations at each temperature, after which the temperature is decreased until the system freezes into a steady state.

Kirkpatrick *et al.* [1] and Cěrny [5] independently showed that the Metropolis algorithm could be applied to optimization problems by mapping the elements of the physical cooling process onto the elements of a combinatorial optimization problem as shown in the following table.

Thermodynamic simulation	Combinatorial optimization
System states	Feasible solutions
Energy	Cost
Change of state	Neighbouring solution
Temperature	Control parameter
Frozen state	Heuristic solution

Thus any local optimization algorithm can be converted into an annealing algorithm by sampling the neighbourhoods randomly and allowing the acceptance of an inferior solution according to the probability given in Equation 2.1. The level of acceptance of uphill moves then depends on the magnitude of the increase in the cost function and on the search time to date.

For most materials the shape of the energy function over the set of feasible states bears little similarity to a typical cost function for a discrete optimization problem, as there tends to be just one very low energy state corresponding to slow cooling. In an optimization problem there are likely to be several local minima which have cost function values close to the optimum. However, materials known as spin-glasses display a reaction called frustration, in which opposing forces cause magnetic fields to line up in different directions, resulting in an energy function with several low energy states. The problems initially considered by Kirkpatrick involved simplified cost functions which took the same form as the energy function for spin-glasses. Thus there is a very close analogy between his initial experiments and the simulation of physical cooling. However, Kirkpatrick extended his work to allow for more general objective functions and, following his success with problems in VLSI design, he conducted some research into the use of simulated annealing in the solution of the well-known travelling salesman problem, showing that the method is applicable to any well-defined objective function. These experimental successes with arbitrary functions have since been backed up by a number of theoretical studies of the statistical processes involved.

The main concern of this chapter is the practical application of simulated annealing to the solution of discrete optimization problems. In order fully to understand the motivation for improving local search methods, and the way in which the annealing algorithm achieves this, it is necessary to be familiar with this type of heuristic technique. Therefore the next section starts with a brief introduction to local optimization followed by a general statement of the simulated annealing heuristic.

As with the traditional descent methods, simulated annealing should not be regarded as a single algorithm, but as a family of algorithms or a heuristic *strategy*. Thus the general statement leaves several decisions which have to be made in order to implement the method for a particular problem. These are covered in detail, and the discussion is backed up by examples and a brief outline of the relevant theoretical results on the convergence of the algorithm. However, no attempt is made to derive or justify these results, and the interested reader is referred to the more detailed texts of van Laarhoven and Aarts [6] and Aarts and Korst [7]. Section 2.3 is concerned with subsequent developments of the basic method. Many of these are the result of extensive experiments in using simulated annealing to solve a particular problem.

While Sections 2.2 and 2.3 include details of applications in order to illustrate or justify the points being made, the full description of practical applications is postponed until Section 2.4. Here an attempt is made to balance the theoretical and investigative work reported in earlier sections by concentrating on practical applications which have been used in the solution of real problems.

2.2 The Basic Method

2.2.1 Local optimization

As simulated annealing is a variant of the more traditional descent methods of local optimization or neighbourhood search, it is useful to start with an outline of this type of heuristic.

Suppose we have a minimization problem over a set of feasible solutions S and a cost function $f : S \rightarrow \mathcal{R}$, which can be calculated for all $s \in S$. In theory, the optimal solution could be obtained by an exhaustive search calculating $f(s)$ for each $s \in S$ and selecting the minimum. However, in most real-life problems the set S will be far too big for this to be practicable. Local optimization overcomes this by searching only a small subset of the solution space. This is achieved by defining a neighbourhood structure on it and searching the neighbourhood of the current solution for an improvement. If there is no neighbour which results in an improvement to the cost function, the current solution is taken as an approximation to the optimum. If an improvement is found, the current solution is replaced by the improvement and the process is repeated. The method of steepest descent searches the whole neighbourhood and selects that neighbour which results in the greatest improvement to the cost function. Random descent selects neighbouring solutions randomly and accepts the first solution which improves the cost function.

Local optimization is a particularly attractive approach for many combinatorial optimization problems as they have a natural neighbourhood structure. If the number of elements in the optimal solution is fixed and known, a neighbourhood can be defined as the set of solutions obtained by swapping a fixed number of elements in the current solution for the same number of non-solution elements. Well-known problems which have been tackled in this way include the travelling salesman problem and the p-median problem.

If k items are swapped, these neighbourhoods are often referred to as k-neighbourhoods, and the solutions obtained by local optimization using the k-neighbourhoods are called k-optimal solutions. There is a trade-off between solution quality and computation time represented by the different values of k. For small values of k the neighbourhoods are small and can be searched efficiently, but the process is likely to converge to a local optimum. For large values of k the neighbourhood size will increase dramatically, but there will be a smaller number of local optima and therefore the chances of attaining the global optimum will be increased.

Thus we can formally state the local optimization process as follows:

Local optimization for a problem with solution space S, cost function f and neighbourhood structure N

Select a starting solution $s_0 \in S$;
Repeat
    Select s such that $f(s) < f(s_0)$ by a suitable method;
    Replace s_0 by s;
Until $f(s) > f(s_0)$ for all $s \in N(s_0)$.
s_0 is the approximation to the optimal solution.
(The usual ways of selecting s are steepest descent or random descent.)

We also note that, even when the number of elements in the solution is not predetermined, a natural neighbourhood structure can be defined by removing or adding an element to the set of solution variables. For example, if the problem is represented by a set of 0-1 variables, a neighbourhood may be defined by changing the value of one or more variables.

2.2.2 The annealing algorithm

As has been suggested, the main disadvantage of local search is its likelihood of finding a local, rather than global, optimum. By allowing some uphill moves in a controlled manner, simulated annealing offers a way of alleviating this problem. The annealing algorithm is similar to the random descent method in that the neighbourhood is sampled at random. It differs in that a neighbour giving rise to an increase in the cost function may be accepted and this acceptance will depend on the control parameter (temperature), and the magnitude of the increase. The algorithm can be stated as follows:

Simulated annealing for a minimization problem with solution space S, objective function f and neighbourhood structure N

Select an initial solution s_0;
Select an initial temperature $t_0 > 0$;
Select a temperature reduction function α;
Repeat
    Repeat
        Randomly select $s \in N(s_0)$;
        $\delta = f(s) - f(s_0)$;
        If $\delta < 0$
        then $s_0 = s$
        else
        generate random x uniformly in the range $(0, 1)$;
            if $x < \exp(-\delta/t)$ then $s_0 = s$;
    Until *iteration_count* = *nrep*
    Set $t = \alpha(t)$;
Until stopping condition = true.
s_0 is the approximation to the optimal solution.

Notice that t is now simply a control parameter and has no physical analogy; Boltzmann's constant has therefore been dropped from the acceptance probability. However it is still usual to refer to t as the 'temperature', and the manner and rate at which t is reduced as the 'cooling schedule'.

The algorithm given above is very general, and a number of decisions must be made in order to implement it for the solution of a particular problem. These can be divided into two categories. Firstly there are generic decisions which are concerned with parameters of the annealing algorithm itself. These include factors such as the initial temperature, the cooling schedule (governed by the parameter *nrep* and the choice of the temperature reduction function α), and the stopping condition. The second class of decisions is problem-specific and involves the choice of the space of feasible solutions, the form of the cost function and the neighbourhood structure employed.

Both types of decision need to be made with care, as they have been shown to affect the speed of the algorithm and the quality of the solutions obtained. Since Kirkpatrick's original paper there has been much research into the theoretical convergence properties of the annealing algorithm. These results show that to guarantee convergence to a global optimum will, in general, require more iterations than an exhaustive search. In spite of this, they do provide pointers as to what factors should be considered in making both generic and problem specific decisions. It will therefore be useful to review some of these theoretical results before embarking on a discussion of the decision making process. This discussion will be illustrated with examples using the travelling salesman and node colouring problems.

2.2.3 A brief overview of the theory

There has been a substantial amount of research into the statistical behaviour of the simulated annealing algorithm. A detailed exposition of this aspect of the method is beyond the scope of this book, and is not necessary in order to be able to use annealing successfully. However, some of the results of this work have proved helpful in determining parameters for the cooling schedules and suitable cost functions and neighbourhood structures. These will be outlined here. The interested reader is referred to the books by van Laarhoven and Aarts [6] and Aarts and Korst [7], both of which give a detailed account of the work on convergence and cite references to many of the original papers in this field.

The theoretical results are based on the fact that the behaviour of the simulated annealing algorithm can be modelled using Markov chains. The probability of moving from one state to another can be represented in matrix form, and for constant temperature the *transition probability* p_{ij} of moving from solution i to solution j depends only on i and j. This type of transition matrix gives rise to what is known as a homogeneous Markov chain. As long as it is possible to find a sequence of exchanges which will transform any solution into any other with non-zero probability, the process converges towards a stationary distribution which is independent of the starting solution. As the temperature tends to zero, so the form of the stationary distribution tends to a uniform distribution over the set of optimal states or solutions.

In simulated annealing the temperature is not constant, but is reduced after a number of iterations. This can be considered either as a number of different homogeneous chains, or as a single non-homogeneous chain in which the transition probability p_{ij} depends not only on the two states i and j but also on the temperature, i.e. the probabilities are dependent on the number of iterations already carried out. Aarts and van Laarhoven [8] show that if a homogeneous chain is to approximate its stationary distribution arbitrarily closely then the number of iterations at a single temperature will be at least quadratic in the size of the solution space. As the solution space is usually exponential in problem size, this means that the running time for simulated annealing with such guarantees of optimality will be exponential.

In the case of the non-homogeneous chain representation, a number of results concerning optimal convergence have been published. The strongest such result to date is that of Hajek [9]. He indicates that the cooling rate depends on the shape of the objective function over the neighbourhood structure. He defines the depth of a local optimum as the increase in cost function needed to escape from that minimum into the valley of any other minimum. A cooling schedule given by $t_k = c/\log(1+k)$, where k is the number of iterations and c is at least as great as the depth of the deepest local (but not global) minimum, will guarantee asymptotic convergence.

These convergence results have given simulated annealing a degree of respectability not always accorded to heuristic methods and there is little doubt that they have helped to generate interest in the technique. However, their usefulness in practical applications is limited, as they imply solution times which are exponential in problem size, and often require more iterations than exhaustive search. In spite of this, they do provide some guidance as to how the different choices presented by the general annealing algorithm should be decided.

2.2.4 Generic decisions

The generic decisions basically involve the cooling schedule, including the upper and lower limits for the temperature parameter and the rate at which it must be reduced. If the final solution is to be independent of the starting solution, the initial temperature must be 'hot' enough to allow an almost free exchange of neighbouring solutions. In some cases there is enough information in the problem to estimate this—for example, if the maximum difference in cost between neighbouring solutions is known, it may be assumed that increases of this magnitude will be sufficient and t can be calculated appropriately. While this may be the case in specially designed test problems, practical applications are often less well-structured and such estimates may be difficult to calculate. In these cases the proportion of accepted moves which represent a suitably volatile system can be decided beforehand, and the system can be heated fairly rapidly until the proportion of accepted moves to rejected moves reaches the required value. At this point cooling can commence. It is interesting to note that this method corresponds to the physical analogy in which a substance is heated quickly to its liquid state before being cooled slowly according to the annealing schedule.

The rate at which the temperature parameter is reduced is vital to the success of any annealing process. This is governed by the number of repetitions at each temperature and the rate at which the temperature is reduced. The theory suggests that the system should be allowed to move very close to its stationary distribution at the current temperature before temperature reduction, and that the temperature should converge gradually to a value of zero. It also suggests that, in order to achieve this, a number of iterations exponential in problem size will be necessary at each temperature. As this tends to imply infeasible computation times some sort of reduction in the number of iterations will be inevitable. This may be achieved either by using a large number of iterations at few temperatures or a small number of iterations at many temperatures.

The two cooling schedules which occur most widely in practice illustrate opposite extremes. The first involves a geometric reduction function $\alpha(t) = at$, where $a < 1$. Experience has shown that relatively high values of a perform best and most reported successes in the literature use values between 0.8 and 0.99, with a bias to the higher end of the range. This corresponds to fairly slow cooling. The number of iterations at each temperature is usually related to the size of the neighbourhoods, or sometimes the solution space, and may vary from temperature to temperature. For example, it is important to spend a long time at lower temperatures to ensure that a local optimum has been fully explored. Thus it can be beneficial to increase the value of *nrep* either geometrically (by multiplying by a factor greater than one), or arithmetically, (by adding a constant factor) at each new temperature.

Another way of determining *nrep* is to use feedback from the annealing process. For example it may be desirable to accept a certain number of moves before decreasing the temperature. This will imply a short time being spent at high temperatures when the rate of acceptances is high. Because the number of acceptances at very low temperatures will be small it may take an infeasible amount of time to reach the required total and it may be necessary to impose a maximum number of iterations per temperature as well. The second commonly-used schedule, first suggested by Lundy and Mees [10], executes just one iteration at each temperature, but reduces temperature very slowly according to the formula $\alpha(t) = t/(1+\beta t)$ where β is a suitably small value.

A number of other cooling rates have been proposed with reference to the theoretical results. For example, the temperature could be reduced according to Hajek's result. However this schedule has not proved popular in practice for the following reasons. The schedule represents a rate of cooling which is so slow as to be infeasible for many problems. The depth of a local optimum is also difficult to estimate and if c is taken to be unnecessarily large this will result in an even slower cooling rate. As an alternative, Aarts and Korst [7] derive a cooling schedule which guarantees a final distribution which closely approximates the required stationary one—a situation they call quasi-equilibrium. The temperature reduction function takes the form:

$$\alpha(t) = \frac{t}{1 + (t \ln(1+\delta)/3\sigma_t)}$$

where σ_t is the standard deviation at temperature t. In spite of its requiring a number of iterations which is polynomial in problem size when combined with appropriate stopping conditions, the cooling rate implied by this schedule tends to be slower than those used in practice.

In theory the temperature should be allowed to decrease to zero before the stopping condition is satisfied. However there is no need to decrease the temperature this far. Given the limited precision of any computer implementation, as t approaches zero the small, but positive, probability of accepting any uphill moves will be indistinguishable from zero. Even before this point is reached, it is likely that the chances of a complete escape from the current local optimum will become negligible. Thus the criterion for stopping can be expressed either in terms of a minimum value of the temperature parameter, or in terms of the 'freezing' of the system at the current solution. Lundy and Mees suggest stopping when

$$t \leq \frac{\epsilon}{\ln[(|S|-1)/\theta]}$$

where S is the solution space.

This is designed to produce a solution which is within ϵ of the optimum with probability θ. Other rules attempt to identify freezing by specifying that a number of iterations or temperatures must have passed without an acceptance. Before this stage has been reached the system may spend time 'cycling' around solutions near to a single local optimum. In order to avoid wasting time in this way the condition is sometimes weakened so that the process stops when the proportion of accepted moves drops below a given value. The simplest rule of all is to pre-specify the total number of iterations and stop when this number has been completed. This rule needs to be carefully tuned with the other parameters to ensure that it corresponds with a sufficiently low temperature to ensure convergence.

Both empirical evidence and the theoretical research suggest that the manner of cooling is not as important as the rate. Thus there is little to choose between, say, a geometric schedule, and the schedule suggested by Lundy and Mees, as long as they cool over the same range of temperatures at approximately the same rate. In view of this result and those of experiments reported in the literature, when using annealing for a new application it is probably best to start off with one of these two schedules first and only consider the others if these fail to provide satisfactory results. In terms of deciding on the values of the parameters for the schedule chosen, there is no easy way of achieving this and almost all the successful applications reported in the literature state that the best parameters were determined after much experimentation.

2.2.5 Problem-specific decisions

The problem-specific decisions are concerned with the solution space, neighbourhood structure and the cost function. Hajek's result supports empirical observations that these factors have a significant effect on the success of an annealing algorithm. As with the generic decisions, it is not possible to set down a series of rules which will always define the best choices for a given problem. However, it is possible to outline some properties which are desirable. Some of these are common sense, some are the result of experimentation, while others are observations stemming from the results on theoretical convergence. In some cases it may not be possible to satisfy all these requirements at once—indeed, some of them may be contradictory for many problem types. In making these decisions there are three important objectives. The validity of the algorithm must be maintained, the computation time should be used effectively for as many iterations as possible, and the solution should be close to the global optimum.

Some of the initial convergence results required that the neighbourhoods should be uniform and symmetric, so that all solutions had the same number of neighbours, and, if solution i was a neighbour of j, then j was a neighbour of i. However, later research showed that the weaker condition that every solution should be *reachable* from every other was sufficient to prove convergence. Thus the first consideration in determining the neighbourhood structure is to ensure that this reachability condition is satisfied. This is usually easy to verify.

If the available computing time is to be used efficiently, it is vital that frequently-used routines should be as fast as possible. An examination of the algorithm shows that two processes are invoked at every iteration. The first is the random generation of a neighbouring feasible solution. While this is a trivial process for many definitions of the solution space and neighbourhood structure, it may prove difficult if the neighbourhoods are large and complex, or if the solution space is constrained by stringent feasibility conditions. As will be illustrated in the example of node colouring below, such restrictions may be overcome by relaxing the feasibility conditions and including a penalty term in the cost function to discourage violations of the relaxed constraints. The difference between the values of the cost function for s_0 and s must also be calculated at every iteration. It is therefore important that the cost function and neighbourhood structure be chosen in such a way that this calculation can be carried out quickly and efficiently. It is often the case that this does not necessitate recalculation of the complete cost function for the new solution, and any such shortcuts should be considered when deciding on the forms of the costs and the neighbourhoods.

If the number of iterations is to be kept reasonably low, Hajek's theorem suggests that it is prudent to avoid neighbourhoods which give rise to a spiky topography over the solution space, and that deep troughs should be avoided if possible. Common sense also dictates that the cost function should be able to lead the annealing process towards the local minima, and thus large plateau-like areas where the cost function takes on equal values should also be avoided.

Several results also demonstrate that the number of iterations depends on the size of the solution space, and this suggests that this should be kept as small as possible. However this recommendation is at odds with the relaxation of stringent constraints suggested above, which will clearly increase the size of the search space by including a number of 'infeasible' solutions. In many instances it may be possible to reduce the size of the solution space by applying a series of reductions to the problem. For example, as will be discussed in detail in chapter 6, the set covering problem has a well-known series of reductions which are normally applied before any solution algorithm is attempted. However, care should be taken that this does not destroy the reachability property. Even if reachability is maintained, such a strategy may have the effect of taking out neighbours which have cost function values close to the current solution, and therefore results in a more jagged topography.

In addition to keeping the solution space small, it is also useful to aim for reasonably small neighbourhoods. This enables a neighbourhood to be searched adequately in fewer iterations, but conversely means that there is less opportunity for dramatic improvements to occur in a single move. Thus there must be some compromise here but, in general, small simple neighbourhoods are preferable to large complex ones.

2.2.6 Examples

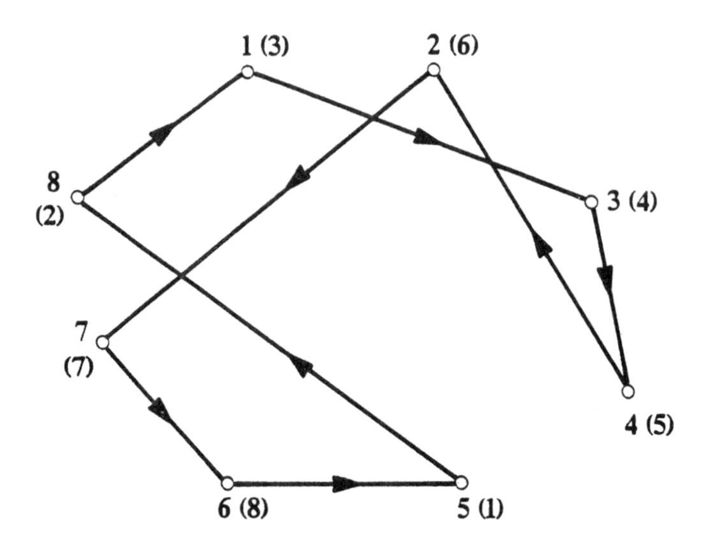

Figure 2.2: An initial TSP tour

Example 2.1 (The travelling salesman problem) *Consider the travelling salesman problem, in which it is required to find a tour of minimum distance passing through a given set of points. We will assume that the required distance measure is the Euclidean or straight line distance.*

Local optimization algorithms for the travelling salesman problem have been widely studied. If the problem consists of n points, any tour can be represented as a permutation of the numbers 1 to n. Physically this corresponds to a circuit through each of the points with links between adjacent points in the permutation and a final link between the nth point and the first. These permutations form the solution space. A k-neighbourhood of any given tour is defined by those tours obtained by removing k links and replacing them by a different set of k links, in such a way as to maintain feasibility of the tour. For $k > 3$, there are a number of ways of reconnecting once k links have been removed. However, if $k = 2$ there is only one way of reconnecting the tour when two of its links have been removed. If we denote by v_i the point in position i of the tour, then if links (v_i, v_{i+1}) and (v_j, v_{j+1}) are removed, the only way to form a new valid tour is to connect v_i to v_j and v_{i+1} to v_{j+1}. For example, Figure 2.2 illustrates a tour on 8 nodes given by the permutation $\{5, 8, 1, 3, 4, 2, 7, 6\}$. The bracketed numbers denote the position of each node in the permutation.

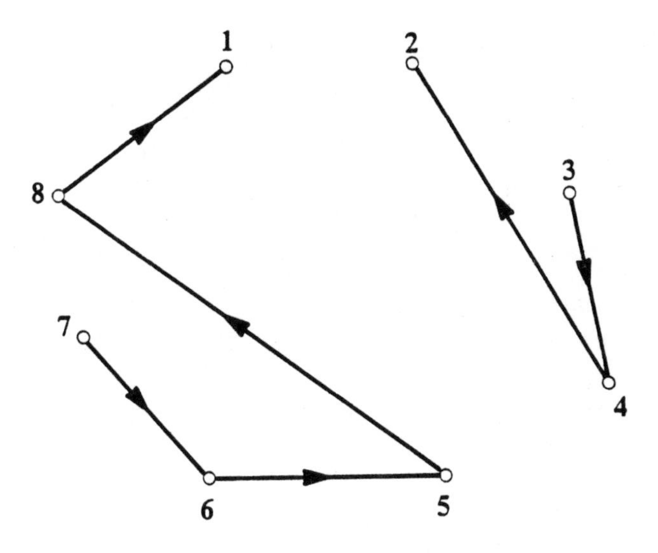

Figure 2.3: After edge removal

If we take $i = 3$ and $j = 6$, links (1,3) and (2,7) would be removed as in Figure 2.3. The only way of reconnecting the 2 resulting sections to form a valid tour is to connect 1 to 2 and 3 to 7, while reversing the direction of travel between 2 and 3, as in Figure 2.4. Thus all 2-neighbours can be completely defined by two indices i and j such that $i < j$.

The size of the solution space is $(n-1)!/2$ (we divide by 2 because our distance function is symmetric), and the size of the neighbourhoods is $n(n-1)/2$. Thus this choice of neighbourhood structure results in neighbourhoods which are many orders of magnitude smaller than the original search space. However it is easy to verify that any tour can be obtained from any other by a sequence of such exchanges.

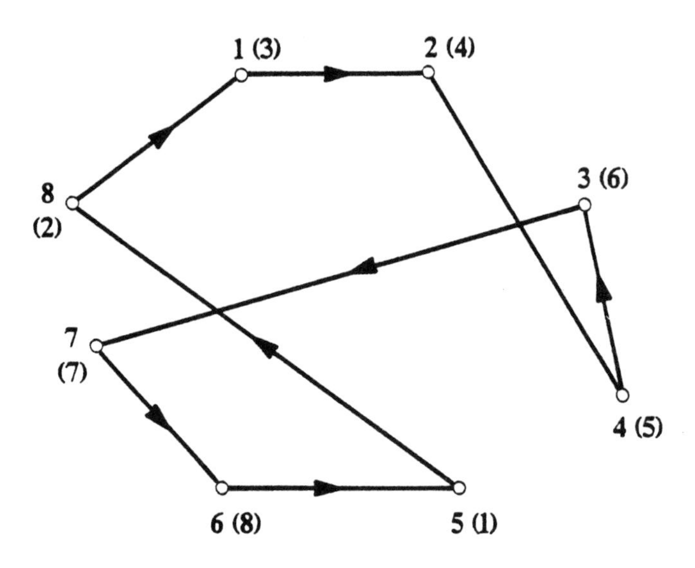

Figure 2.4: Neighbouring tour

It also satisfies the requirement that a neighbouring solution can be easily generated at random by simply generating two uniform random numbers to represent i and j. The obvious cost function is the length of the tour. In order to calculate the increase in cost δ, it is not necessary to calculate the length of the new tour from scratch. δ is obviously given by:

$$d(v_i, v_j) + d(v_{i+1}, v_{j+1}) - d(v_i, v_{i+1}) - d(v_j, v_{j+1}).$$

Thus this combination of the definition of feasible solutions, neighbourhood structure and cost function appears to be a good candidate for annealing. A starting solution can be obtained by generating a random permutation of the indices and the algorithm can then be applied directly using the chosen cooling schedule.

Example 2.2 (Node colouring) *The node colouring problem is a well-known problem in graph-theory which has many applications, particularly in time-tabling and scheduling. A graph can be defined as a set of points or nodes, some of which are connected to each other by lines or edges. Two nodes connected by an edge are said to be adjacent. The node colouring problem is concerned with allocating a 'colour' to each node such that adjacent nodes are not given the same* *colour. Figure 2.5 shows a valid 4-colouring with colours A, B, C and D. The objective is to find an allocation which uses the smallest possible number of colours.*

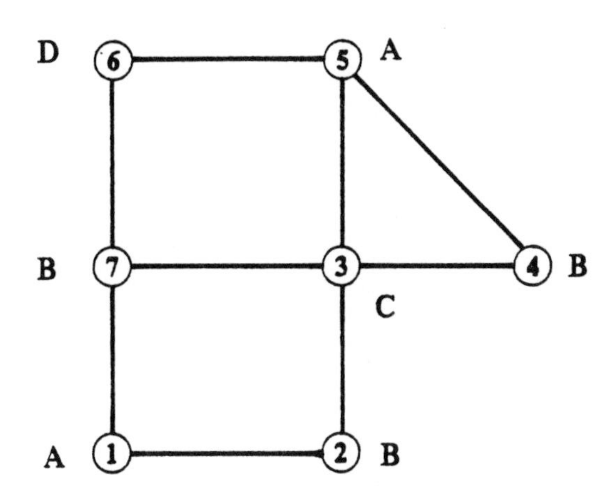

Figure 2.5: Graph with 4-colouring

The node colouring problem differs from the travelling salesman problem in that it has not previously been viewed as a candidate for local optimization algorithms. However several researchers have attempted to solve the problem using an annealing approach. The natural definition of the solution space is the set of feasible colourings. These can be regarded as partitions of the nodes into k subsets such that there are no edges between any two nodes in the same subset. All nodes within a subset can then be coloured in a single colour, so the objective is to minimize the number of subsets used. The subsets for the colouring in Figure 2.5 are $\{1,5\}, \{2,4,7\}, \{3\}$ and $\{6\}$.

The neighbourhood of a given solution could be defined by swapping some of the nodes between 2 subsets. The problem here is that this must be done in such a way as to maintain feasibility. Arbitrary swaps are likely to introduce edges into the subsets leading to solutions which fall outside the solution space. In the example it is not feasible to swap nodes 1 and 2, as node 1 cannot be in the same subset as node 7. This problem can be overcome by using Kempe chains. If we consider only that part of the graph defined by the nodes in any 2 of the subsets, then it will contain a number of components which are completely disconnected from one another. (There may be just one component). These components are the Kempe chains. Figure 2.6 shows the AB-Kempe chains for the colouring in Figure 2.5.

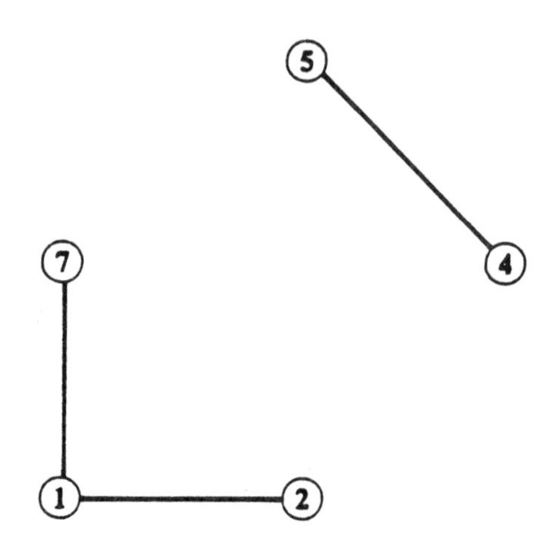

Figure 2.6: AB-Kempe chains

If any node from a given chain is moved from its current subset into the other, then to maintain feasibility, all the other nodes in the chain will also have to swap subsets. This is illustrated in Figure 2.7. Thus feasibility can be maintained by defining the neighbouring solutions as those obtained by swapping the colours of nodes in a single Kempe chain. However, this can only be achieved at the expense of calculating Kempe chains at every iteration, or by storing and updating all the chains throughout the annealing process. Thus this neighbourhood is not an ideal candidate for the annealing algorithm, although careful implementation can minimize the effect of these computational difficulties, especially for dense graphs where the sizes of the subsets tend to be small.

An alternative approach is to relax the definition of feasibility, and include a penalty in the cost function to ensure that infeasible solutions are expensive. The feasible solutions can now be defined as any partition of the nodes into subsets, and a term involving the product of a penalty factor and the total number of edges between nodes in the same subset can be added to the cost function. Neighbouring solutions can now be generated simply by moving a node

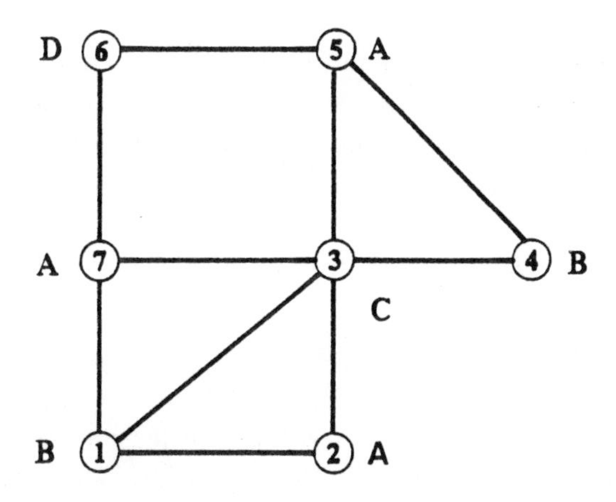

Figure 2.7: Valid colouring after swapping on chain (7,1,2)

from one subset to another.

The natural cost function is the number of subsets in the partition. This also provides problems for an annealing process. The value of the objective will not change unless one of the subsets becomes empty. Thus we have the undesirable situation of large plateau-like areas in which exchanges will be accepted freely, and the algorithm is likely to wander around aimlessly with no guidance towards improved solutions. Although the inclusion of a penalty for infeasibility will introduce variation, this will not guide the annealing in the direction of fewer subsets or colours. What is required is a way of indicating that solutions in which there are a number of large subsets balanced by some very small sets, is an improvement on a solution in which the nodes are partitioned fairly uniformly.

Aarts and Korst [7] give a cost function which is based on a series of penalty factors. The colour sets are indexed and the penalties are derived in such a way as to favour large, lower indexed sets, thus encouraging the higher indexed sets to shrink and eventually disappear. However, they report that these penalties can get very large for realistically sized problems, which leads to computational difficulties. To overcome this they suggest a relaxed set of constraints on the penalties which allows a less rapidly increasing series of values, but does not guarantee that the optimal solution will correspond to the minimum number of colours. Johnson *et al.* [11] suggest an alternative cost function made up of a penalty term for infeasible colourings, and a term to encourage large subsets. They do not include the number of sets explicitly, but rely on the large subset term to produce this as a side effect. The form of their function is:

$$-\sum_{i=1}^{k} |C_i|^2 + \sum_{i=1}^{k} 2|C_i||E_i|$$

where $|C_i|$ is the number of nodes, and $|E_i|$ the number of edges, in subset i. This function does not guarantee a minimum number of colours but the local minima will all be feasible colourings. The ability of the annealing algorithm to visit several local minima means that an optimal colouring may well be found during the search. Johnson *et al.* present a series of very thorough experiments using this objective with both the Kempe chain neighbourhoods and the relaxed/penalty version, and report that the Kempe chain approach is slower but can produce better results.

It is worth noting that for some applications it is not necessary to minimize the number of colours, but to find a feasible colouring within a given number of colours. In this case the solution space can be defined as the set of partitions of the nodes into the required number of subsets, and the penalty function alone can be used to drive the process to the optimal cost of zero—representing a feasible solution. See Chams *et al.* [12] and Johnson *et al.* [11] for further details of this type of approach.

2.2.7 Aids to fine-tuning

The above discussion shows that while there are many pointers to aid in the decisions concerning both the generic and problem specific parameters, there will still be a number of choices which can only be finalized by experimentation on typical problem instances. The design and analysis of such experiments is complicated by the fact that annealing is a randomized algorithm, and therefore the results of a single run may not be typical. In this section we summarize some of the techniques which have proved useful to such analysis.

There is little doubt that pictorial representations are invaluable in understanding the evolution of an annealing run. For some problems it is possible to show the annealing process at work. For example a number of researchers have gained valuable insight into the performance of their annealing implementations of the travelling salesman problem by watching the tours unfold on the computer screen. Dowsland [13] reports that a graphical implementation of an annealing process for two dimensional packing problems was useful in identifying those decisions which were worth investigating by more rigorous testing, and those which could be abandoned straight away. Information concerning relevant parameters such as temperature, the value of the cost function, and the ratio of accepted moves to rejected moves all add to the usefulness of such displays, and the insight gained is almost certainly worth the extra effort required in producing the graphics routines. However a word of warning is due here. Watching the annealing process unfolding before you can be addictive both to yourself and any one who happens to be passing! On the plus side this can generate interest in your work from unexpected quarters, which in turn fuels new ideas for algorithm improvement.

Even if an application is not amenable to graphical representation, some form of graph or plot of the algorithm's progress can be achieved. Valuable information can be gleaned from graphs of cost function, percentage accepted moves etc., plotted against time or number of iterations. Plots highlighting the effects of different experimental parameters such as different starting temperatures, stopping conditions etc. often display trends not immediately noticeable from numeric data.

Three papers by Johnson *et al.* [11, 14, 15], which cover a wide range of experiments comparing annealing with other heuristic methods for a number of classical problems, provide an excellent background in the design and analysis of annealing experiments. One particularly interesting technique used to determine suitable values for many of the parameters is the application of box plots. These allow a coarse measurement (in quartiles) of the effect of the parameters on important factors such as solution quality and running times. This is illustrated on a two-dimensional plot which allows the analyst to see at a glance those values of the parameter which are good with respect to the relevant factor, and also how sensitive the results are to this choice.

Many researchers, especially those involved in the earlier experiments with the simulated annealing heuristic, suggest that the analogies to physical quantities such as entropy and specific heat may provide clues as to the efficiency of a particular schedule. Entropy is a measure of the disorder of the system given by

$$-\sum_{i \in S} p_i(t) \ln p_i(t)$$

where $p_i(t)$ is the probability that the system is in state i at temperature t. Entropy should be monotonically decreasing as temperature decreases.

It has also been suggested that there may be a peak in specific heat—the rate of change of mean energy—as the point is approached where very slow cooling is needed. While there are some experimental results which demonstrate how closely the observed estimates of such measures are related to those predicted by the theory, many recent researchers report that they have not found them useful, and they do require a further overhead of calculation. While such calculations may provide some insight into a particular problem they have no real meaning in an optimization context. Thus, if the basic algorithm fails to produce the required results after a reasonable amount of experimentation it is probably more worthwhile to move further away from the physical analogy by trying some of the modifications described in the next section.

2.3 Enhancements and Modifications

The annealing algorithm described in the last section closely follows the original Metropolis simulation. In doing so, it maintains its analogy with the physical cooling process, and also satisfies all the conditions required for optimal convergence given a suitable cooling schedule. As we have seen, the run-times of schedules which can be *guaranteed* to approach the stationary distribution of optimal values are usually prohibitively long. Thus it is arguable that an algorithm which works in the spirit of the annealing method—controlling uphill moves by a probability function based on temperature—but which does not adhere so closely to the original simulation, may well prove effective. The convergence proofs are known to hold under less restrictive conditions, and even if the convergence conditions are violated the algorithm may perform well in practice. In this section we examine a number of modifications which have proved useful in adapting the annealing algorithm for a number of different problems. These have been classified under a number of different headings indicating that part of the annealing algorithm to which they are applied.

2.3.1 Acceptance probability

Perhaps the most obvious modification in moving away from the physical analogy is to consider changing the probability of accepting an uphill move. The use of the Boltzmann distribution is entirely due to the laws of thermodynamics, and there is no reason to suppose that some other distribution would not perform better for certain problems. However, there are relatively few examples of changes in distribution to be found in the literature. This is no doubt due to the fact that the Boltzmann distribution does exhibit characteristics which are intuitively correct for the acceptance criteria. Solutions are accepted according to the magnitude of the increase, but the distribution is not linear, so that large increases have virtually no chance of acceptance whereas small increases may be accepted regularly. However, some authors have reported success with different distributions.

There appear to be two reasons for experimenting with different acceptance probabilities. The obvious reason is a lack of success with the usual exponential criterion. Less obvious, but of more widespread importance, is the problem of algorithm speed. Johnson *et al.* [14] found that the calculation of $\exp(-\delta/t)$ at every iteration accounted for about one third of the total computation time in an annealing solution to a graph partitioning problem. They suggest two possible methods of improvement. The first is to use $p(\delta) = 1 - \delta/t$, which approximates the exponential. However, a better approach is to use a discrete approximation represented by a look-up table which can be calculated at a series of fixed values over the range of possible values for δ/t. Johnson *et al.* used all integer values in the range. The approximation was then obtained by simply rounding δ/t to the nearest integer and looking up the appropriate function value. An implementation of this modification succeeded in reducing the running time by about one third and did not exhibit any adverse effects on solution quality.

Other researchers have found that the standard exponential distribution does not perform well for their particular problem and have found other distributions more successful. For example, Dowsland [13] considers a two dimensional packing problem in which the objective is to fit a given set of small rectangles without overlap, into a larger containing rectangle. The solution space is made up of all possible placements, and the cost function is a measure of the overlap which must obviously be reduced to zero. A neighbourhood is defined by a movement of any single piece. Analysis of the reasons for poor annealing performance with the standard acceptance function for this problem identified a need to accept moderate increases in the cost function in order to escape from local minima. However, when the temperature was high enough for such increases to be accepted with non-negligible probabilities, small increases of one or two units were accepted with far too high a probability. The best way of dealing with this proved to be the addition of a constant to the value of δ for all $\delta > 0$, which is equivalent to moving to a less steep region of the exponential curve.

Others have found that simpler functions can give good results. Such simple functions are particularly prevalent in problems in production scheduling—for example, Brandimarte *et al.* [16] use a probability of the form $p(\delta) = \delta/t$, and Ogbu and Smith [17] and Vakharia and Chang [18] both use probabilities which are independent of δ for different sequencing problems. It should be noted that such changes in the probability function may result in a situation where $p(0) \neq 1$. The decision as to whether to accept moves for which $\delta = 0$ automatically, or according to the calculated probability then becomes yet another parameter to be decided by the algorithm designer.

2.3.2 Cooling

Examination of the solutions produced throughout the cooling process for almost any problem will reveal that most of the useful work is done in the middle of the schedule. Starting the process with the temperature so high that almost all exchanges are accepted simply produces a series of random solutions, each one of which might itself have been a starting solution. This part of the search also tends to be slower, as the large number of acceptances means that a large number of solution updates must be executed. Some researchers suggest that cooling should be very rapid during this phase, and achieve this by reducing temperature after a fixed number of acceptances, if these are achieved before the iterations count reaches the *nrep* value. Others suggest starting at a lower temperature at which the rate of acceptance is relatively small.

Conversely, when the temperature is very low, little movement is possible and, as few increases in the objective are accepted, the process degenerates into a very slow version of the descent algorithm. Thus it may be advantageous to search only the middle part of the temperature range, using the time saved to allow slower cooling. Connolly [19] was probably the first to suggest taking this to the extreme of annealing at a constant temperature. Such a temperature must obviously be high enough to allow the process to climb out of local optima, but cool enough to ensure that these local optima are visited.

The problem with this approach is that a good temperature will not only vary from problem type to problem type, but will vary for different instances of the same problem. Therefore, it is necessary to have a robust method for determining a good temperature for a given problem. It has been suggested that specific heat will be high at the point where careful cooling becomes necessary. Connolly experimented with this idea in an attempt to identify the best temperatures for solving quadratic assignment problems. He reports that this was unsuccessful. Instead he suggests another way of determining the optimum temperature. He defines a wide range of temperatures which encompass the range in which the optimal temperature is expected to lie. Then he defines a number of iterations which will be set aside for temperature determination. (The rest of the time will be spent at this temperature). The first pass uses the cooling schedule of Lundy and Mees [10] in which there is just one iteration at each temperature and cooling takes place according to $t \to t/(1+\beta t)$. If K iterations are required to anneal from t_s to t_f then this can be achieved by setting

$$\beta = (t_s - t_f)/K t_f t_s.$$

The temperature which gives the minimum valued cost function during this cooling phase is recorded; this is then used as the constant temperature for the remainder of the algorithm.

If we accept that there is no reason to cool when applying simulated annealing to an optimization problem, then we may also sanction the use of heating. Glover and Greenberg [20] suggest that this approach mirrors more closely human behaviour in trying to find the lowest point in a range of hills. One's inclination is to move downhill, but if no progress is apparent in searching the current valley a concerted uphill effort would be made in order to widen the scope of the search. In fact Kirkpatrick *et al.* [1] used reheating in their original experiments (although this is only explained in the original detailed version of their report). They used the algorithm in an interactive manner, watching the travelling salesman tours unfold and triggering a reheat when the process got stuck in a configuration which displayed obviously sub-optimal features. Such interaction is not always possible, but reheating could be triggered automatically by factors such as the rate of change in the cost function or the ratio of acceptances.

Dowsland [13] suggests a more gradual process, designed to reproduce the effects of single temperature annealing on a spiky objective function, which meant that it was impossible to find a temperature which tended to converge to local optima and still be able to escape from them. Every time a move is accepted the system cools according to the function $t \rightarrow t/(1+\beta t)$, and every time a move is rejected the system is heated according to $t \rightarrow t/(1-\gamma t)$. If $\beta = k\gamma$ then the system will need to go through k heating iterations to balance one cooling. If the ratio of rejected moves to accepted moves is greater than k the system heats up. If it is less the system cools. Thus this schedule theoretically tends to converge to a situation in which the ratio is about k. As it is important adequately to search the areas close to local optima without a significant temperature increase, it is suggested that k should be governed by the size of the neighbourhoods around these minima. This schedule was found to work well on the rectangle packing problem where both systematic cooling and a constant temperature approach had failed to produce results.

2.3.3 The neighbourhoods

Implicit in the statement of the algorithm as given in the previous section is the assumption that the neighbourhood structure is well-defined and unchanging throughout the algorithm. In practice this need not be the case, and improvements in performance can be obtained by adjusting the neighbourhood structure as the temperature decreases. This usually takes the form of restricting the neighbourhood in some way. An example is given in an annealing heuristic for the placement phase of VLSI design, in which rectangular blocks are placed on the chip area in such a way as to minimize a combination of cost factors, described by Sechen *et al.* [21]. They include the horizontal and vertical translations of any block in the set of valid neighbourhood moves. As only small translations tend to be accepted at low temperatures, much time is wasted generating and rejecting longer translations. In order to avoid this, a limit on the maximum translation length is imposed and this is decreased as the temperature drops.

In situations where a penalty function is used to enforce constraints, the neighbourhood size can be decreased by allowing only moves involving variables which contribute to the violation of constraints. For example, in the rectangle packing problem, Dowsland found it beneficial to move only pieces which contributed to the overlap. In some situations this type of restriction may destroy the property that all feasible solutions are reachable from all others, and thus violate the conditions of the convergence proofs. As long as the optimal solution remains reachable from all others which are themselves still reachable, this may not be detrimental to the performance. However Tovey [22] suggests that a better performance may be achieved if the reduced neighbourhood is used with a fixed probability, and the full neighbourhood is used for the remaining iterations. In this way the convergence conditions are maintained and, although the algorithm is biased in favour of removing constraint violating elements, other moves are still feasible.

Some success has also been reported using neighbourhood structures which could potentially destroy reachability. In the paper by Rossier *et al.* [23], for example, the size of the 2-opt neighbourhoods for the travelling salesman problem is reduced by restricting the two generating points i and j to those which are 'close' to one another. Various definitions of closeness (e.g. horizontal or vertical distance, Euclidean distance) are used, and empirical evidence suggests that the smaller (closer) neighbourhoods yield the best results. Problems in which the points are highly clustered will certainly not satisfy the reachability conditions under such a neighbourhood structure. However it is not clear whether the problems used for Rossier's experiments maintained reachability or not.

2.3.4 Sampling

The standard annealing process samples randomly from the neighbourhood of solutions. It is clear that when the process is close to a local optimum most of the neighbouring solutions will involve increases in cost. With random sampling it is possible that some uphill move will be accepted before the local optimum is sampled, and thus the local optimum may never be reached. In order to lessen the chances of such an occurrence it has been suggested that sampling should be cyclic rather than random, to ensure that all neighbours are tried once before any are considered for a second time. This also has the added advantage of avoiding the need for determining a random neighbour. As random number generation is not a computationally trivial task this can result in some speed up of the algorithm. However it is important to impose a random ordering on the cycle at the start of the algorithm to avoid any natural ordering of the solutions affecting the randomness of the process.

Consider, for example, the quadratic assignment problem in which a set of n items is to be assigned to n locations in such a way that each item is at a single location, and each location contains a single item. The cost function is a sum of linear and quadratic terms. A natural neighbourhood structure for such a problem is defined by swapping the locations of two items. The natural ordering would take all swaps for either the first item or the first location before moving on to the second. Sampling in this order is likely to affect the randomness of the search and therefore a random permutation scheme should be used on the natural ordering before sampling commences. Connolly [19] reports that this cyclic sampling improved the performance of annealing applied to the quadratic assignment problem and other researchers report similar success for other problems. It should however be noted that there are also reports of cyclic sampling having an adverse effect on the solution quality for some annealing implementations.

Towards the end of the search when the system is relatively cool much time is spent evaluating moves which are rejected. This can sometimes be avoided by determining the acceptance probability for each move in the neighbourhood, sampling the neighbourhood using a weighted distribution given by these probabilities and accepting automatically. Greene and Supowit [24] show that this is equivalent to annealing in the usual way. They use this approach for the graph partitioning problem in which the nodes of a graph are to be partitioned into 2 equal subsets in such a way that the number of edges connecting nodes in different sets is minimized. The neighbourhoods are defined by moving a single node from one set to the other, any imbalance between the sets being penalised in the cost function. Using this cost function and neighbourhood structure it is easy to keep track of the probabilities needed for the calculations.

For other cost functions this may not be such a simple calculation, and this modification will not generally be worthwhile if the whole neighbourhood must be searched and evaluated at each iteration. However, at local minima and low temperatures, if cyclic sampling is used, the whole neighbourhood may be searched without an acceptance. In this case a move could then be selected according to the above rule as all the probabilities will have been calculated as a part of the sampling process. It should be noted that this may well interfere with the stopping criterion adopted, which may have to be modified accordingly.

2.3.5 The cost function

In some instances it may not be possible to choose a cost function for which the difference between the current and new solutions can be calculated quickly. Tovey [22] suggests that an approximation may be used. If the correct solution is calculated occasionally in addition to the approximation then adjustments in the acceptance probabilities can be used to allow for the error in the estimate. He quotes a simple example in which $t = 10$ and the true change in objective is 5 whereas the approximation suggests a value of 8. The correct acceptance probability is $\exp(-5/10) = 0.61$, whereas the estimate suggests $\exp(-8/10) = 0.45$. If the probability of using the correct function is 1/3 and the probability of using just the estimate is therefore 2/3, this solution would be accepted with probability 0.45 two thirds of the time. In order to attain the correct overall probability of 0.61 it is necessary to accept with probability 0.93 for the remaining third as $(2/3)(0.45)+1/3(0.93) = 0.61$. This technique can be used to adjust the probabilities on the occasions when both values are known. The proportion of the iterations which are calculated accurately will depend on the expected accuracy of the approximation.

Even without such an adjustment strategy, it is possible to obtain good results with a cost function which does not precisely represent the true objective. If the true objective is evaluated only for each accepted move then the true minimum out of all those visited can be retained. This is the situation with the colouring problem discussed in the last section. It is a particularly useful approach to many complex problems in which feasibility is controlled by means of a penalty function, as low valued feasible solutions can be identified even though they may not correspond to global or even local optima with respect to the cost function governing the annealing process.

The importance of a cost function which guides the annealing process towards local optima has already been discussed. Even if this is achieved, a system at a saddle-point between 2 valleys will move to either with equal probability. If one choice leads to the global and the other to a local optimum, a move in the wrong direction may never be recovered. Sometimes this is unavoidable, but in other cases a change in the form of the objective may highlight the one direction as an improvement and the other as a hill-climbing move.

This situation was encountered by Dowsland for the rectangle packing problem. In this case a relaxation of the cost function identified the correct move as downhill and the inferior move as uphill. However, the relaxed objective alone was insufficient to converge towards 'feasible' solutions with no overlap. One way of using the properties of both objectives is to express cost as a weighted sum of the actual and relaxed costs. Dowsland reports that this did not work well; better results were obtained by annealing with respect to both functions in parallel and only accepting moves which were acceptable for both functions. The same temperature parameter was used for both acceptance functions, but multiplied by different weighting factors for the 2 different cost functions. The best results were achieved when the acceptance probability distributions were different—the one being exponential and the other linear.

Where the cost function involves a penalty factor it can be difficult to determine the correct weighting factors for the different terms. We have seen that for problems such as graph colouring it is important to use the right balance between the different parts of the function. In addition, where the true cost function takes on relatively few integer values, it is often necessary to use other cost elements to guide the annealing process across the resulting plateau-like areas. One way of avoiding these problems is to solve the problem iteratively, trying to attain feasibility for decreasing values of the true cost function. In this way only the penalty function is involved in the annealing cost, and thus no weighting decisions are required. The solution space may also be reduced as it contains only those solutions which achieve the current (constant) true cost. For example, in the graph colouring problem repeated applications of the Chams *et al.* [12] approach would be used to find solutions using k colours for a decreasing set of values k. In a similar way, Dowsland tackles the problem of minimizing the length of material required to pack a given set of rectangles as a series of problems attempting to find a feasible packing of the pieces into a fixed length rectangle. Once a feasible solution is found the length is reduced.

The main disadvantage with this approach is that a globally optimal solution with a cost value of zero must be found at each stage in order to attain feasibility. In many situations this may be expecting too much of the annealing algorithm. On the other hand, the optimal value is known and therefore time is not wasted searching for a better solution after the optimum has been achieved. Johnson *et al.* [11] report that, for the colouring problem, this approach out-performs the Kempe-chain and penalty function approaches for graphs for which the number of nodes in different colour classes are about equal. This is because the objective function used for these methods rewards unbalanced classes. Dowsland also reports success with this approach for the packing problem in which annealing generally out-performed other well-known heuristics.

2.3.6 In combination with other methods

While simulated annealing has been shown to be a useful heuristic technique for a variety of problems, in some instances its performance can be improved considerably by combining it with other heuristic methods. In general such approaches are confined to some form of pre-processing before the annealing process is invoked, or post-processing in which attempts are made to improve solutions encountered by the annealing algorithm. However there are examples of heuristics being used as a part of the annealing algorithm. One such application is reported in the next section.

The most obvious use of a pre-processing heuristic is in determining a good starting solution instead of generating one at random. If this approach is adopted, starting the run at a high temperature will effectively destroy the characteristics of this good solution. The search must therefore be commenced at a lower temperature. This has the advantage of saving a substantial amount of solution time. However it may not be advisable to start with too good a solution, especially if the initial temperature is low, as the process may never fully escape from the neighbourhood of the starting state. Nevertheless, many successful implementations report using a starting solution which is better than totally random.

Another way of incorporating some knowledge into the starting solution is to pre-define some of its features. These could theoretically be destroyed during the annealing process but this is unlikely at lower temperatures. An example of this is the approach taken by Bonomi and Lutton [25] for the travelling salesman problem. They divide the area containing the locations into a set of regions and find a good starting tour through the centres of the regions. Locations within each region are ordered randomly but the regions are ordered according to this initial tour. Empirical evidence suggests that this does improve on a completely random starting point, but there are obviously situations in which the ordering of the regions could be specified wrongly and this may never be corrected during the subsequent iterations.

A third approach is to incorporate annealing into a construction heuristic which works by building onto a previous partial solution. This approach was used in [12] by Chams *et al.* in their colouring algorithm. One of the best colouring heuristics works by selecting the colour groups one colour at a time in an intelligent way. It is likely that decisions made early in the process are good decisions and will not adversely affect future allocations. However there comes a critical point where the wrong decisions may be made thus leaving the remaining nodes requiring more than the optimal number of colour classes. Chams *et al.* suggest colouring a (large) proportion of nodes using the heuristic, then using annealing in an attempt to minimize the number of colour classes for the remainder by iterating through different values of k. They report that this combination is more successful then either the construction method or annealing used alone.

The most obvious form of post-processing is to apply a descent algorithm to the final annealing solution to ensure that at least a local minimum has been found. Some researchers suggest applying a descent phase more frequently, but it is difficult to determine when this should be done. One extreme is to apply it after every accepted move. This was attempted in a production scheduling application, but not surprisingly the resulting algorithm proved too slow—several hundred times slower than an equivalent cooling schedule without the descent phase.

For some problems there may be simple transformations which would attain the optimal solution, but which do not fall into the definition of legal neighbourhood moves. This is especially true where local optima may correspond to infeasible solutions which could be made feasible with no increase in cost. In such situations it can prove useful to call an exhaustive search using a different definition of neighbourhood. For example Dowsland reports that the annealing process for the packing problem often passed through phases in which just two pieces overlapped and it was obvious that a swap in the position of one of these with a non-overlapping piece would find a feasible and therefore optimal solution. Such swaps are generally too costly to compute to form a regular neighbourhood structure, but calling an exhaustive search for such swaps whenever just two pieces overlapped proved very worthwhile. A graphical representation of the annealing process revealed other situations where the path to an optimal solution was obvious but did not comprise a set of legal moves, and it is possible that further research into this aspect of the problem could prove very fruitful.

2.3.7 Parallel implementations

There is ever increasing interest in the use of parallel processing in order to attain greater speed, and it is not surprising that the suitability of the annealing process for parallelization has been the subject of much recent research. This is especially so in applications such as image processing and VLSI design.

Aarts and Korst [7] identify three ways in which parallelism may be introduced into the annealing process. The simplest is to allow different processors to proceed with annealing using different streams of random numbers until the temperature is about to be reduced.

At this point, the best result from all the processors is chosen and all processors recommence at the new temperature starting from this common solution. This will result in significantly different chains of solutions when the temperature is high, but when the temperature is lower and there is less flexibility it is likely that the processors will end up with solutions which are close in terms of the neighbourhood structure and cost.

A second strategy is to use the processors to generate random neighbours and test for acceptance independently. Once one processor finds a neighbour to accept then this is conveyed to all the other processors and the search moves to the neighbourhood of the new current solution. This strategy will be wasteful at high temperatures when almost all solutions are accepted, as many of the processors are likely to find an acceptable solution but only one will be used. Conversely at low temperatures when the majority of solutions are rejected this will speed up the search considerably. Thus the best solution would appear to be to start with the first strategy until the ratio of rejections to acceptances exceeds a certain level and then to switch to the second strategy.

Finally it is possible to hold the current solution in common memory and to allow all the processors to act on it independently, each one generating a neighbour and updating the solution if the neighbour is accepted. However, it is possible that this will result in two moves being made which both give an improvement when considered independently but result in an uphill move if they are both carried out. This approach has been used successfully for a school timetabling problem by Abramson [26]. He defines a neighbouring solution as one obtained by moving a class from one time slot to another. The objective is made up of the weighted sum of penalties representing different forms of clashes and feasibility violations. The contribution of a class to the penalty depends only on the other classes in the same time-slot. When a processor considers a move the two time-slots involved are locked from the other processors. Thus all the classes which are involved in the resulting change in the cost function are also locked. The other processors will be working on unrelated time-slots, so that the changes in cost they calculate will be independent of those calculated by any other and the problems described above will not arise. This locking process does have an overhead and Abramson reports that for a given number of processors there is a threshold value on the number of time-slots below which it is better to use a sequential approach.

The wide variety of modifications described in this section illustrate the vast amount of research which has been carried out since Kirkpatrick's original paper. Almost all recently published work deviates from the basic annealing algorithm in some way. Certainly these modifications have improved the performance of annealing on the problems to which they were applied. However, it is possible that the basic algorithm has been applied successfully to a variety of problems not reported in the literature.

The great variation in approaches described here presents the algorithm designer with considerable problems in deciding which, if any, of these may prove profitable for his particular problem. As they mostly involve further decisions concerning parameters it is clear that they cannot all be investigated thoroughly for a given problem. However experience with the basic algorithm over a variety of parameter settings often provides a significant insight into those areas which may be performing poorly, and the most beneficial modifications can then be identified.

The search for improved annealing schedules is an on-going process limited only by the algorithm designer's imagination, and by the the basic ground rules of a local search with probabilistic acceptance of uphill moves governed by a control parameter. Experiments on a particular problem may well suggest modifications other than those outlined here.

2.4 Applications

This section is concerned with the reported applications of simulated annealing to a variety of problems. The emphasis is on the solution of practical problems, and a number of problem areas where annealing is emerging as a useful approach will be identified. Before discussing some of the more applications oriented implementations, we will take a brief look at the success of annealing for a number of classical problems in combinatorial optimization and graph-theory.

2.4.1 Classical problems

There have been a number of independent studies investigating the use of the annealing algorithm in the solution of a variety of classical combinatorial optimization problems. As most of the studies use different cooling schedules, a variety of different enhancements and modifications, and are coded in different languages and run on different machines, it is difficult to draw any definite conclusions concerning the quality of an annealing approach when compared with other heuristics or with a different annealing schedule.

Some of the questions thus posed have been answered by a remarkably thorough investigation carried out by Johnson *et al.* and reported in 3 parts, [11, 14, 15]. These articles cover the graph partitioning, graph colouring, number partitioning and travelling salesman problems. They present results and analysis of the performance of a variety of annealing schedules and neighbourhood structures which can be applied to these problems, and compare them with each other and with the best known heuristics. In particular the results of annealing are compared with those obtained by spending the same amount of time running a descent algorithm using different randomly generated starting points. The random data sets of problems used for the tests were generated from different distributions, so that the performance could also be analysed with respect to different features of the data. In addition these articles provide comprehensive surveys of the use of annealing for the problems in question. They also describe in full the design and analysis of the experiments carried out to pinpoint good parameter values for each problem type, and are invaluable reading for anyone planning serious involvement with the technique.

Other classical problems not studied in this series but considered elsewhere are the quadratic assignment [19] and Steiner [27] problems. Although the detailed conclusions vary from problem to problem and from paper to paper there appears to be some general agreement. In order to obtain good solutions annealing does require a significant amount of time, and within reason it is better to use this time cooling slowly in a single run rather than restarting several shorter runs. However, if a very long period of running time is available it may be advantageous to allow several restarts. The decision as to how to divide the time will obviously depend on the problem size and type and the annealing parameters used, and is yet another variable to be decided empirically.

Almost all those who have made the comparison report that simulated annealing gives better results on average than repeated applications of a straightforward descent algorithm over the same neighbourhood structure. However, for problems such as the travelling salesman problem which have been well researched, specialist algorithms can outperform annealing by a considerable margin. For many problems annealing performs better on uniform data than on data which is clustered in some way. This may be due to the smoother topography generated by the uniform problems and it is possible that allowing some form of reheating may improve the performance for these clustered problems, by allowing more frequent escapes from deep troughs.

2.4.2 VLSI and computer design

The design of electronic systems is a complex one involving many, often conflicting, goals. For example a major factor in response times is the length of the wiring, but the shortest routes often lead to a degree of congestion which cannot be accommodated. The problem involves partitioning, placement and routing decisions which will all interact and have an effect on the optimality of the final design. The complexity of the problem means that these three aspects are usually solved separately, but the resulting subproblems are still extremely complex. This, together with the great variation in the requirements of different designs, means that traditional heuristic methods are unable to provide flexible solutions.

Kirkpatrick *et al.* [1] showed how annealing could be used in the solution of all three phases of the problem. The partitioning phase can be modelled directly as the graph partitioning problem discussed above. The placement problems vary from those in which the components must be placed on a grid, to those in which a variety of different sized components have complete freedom of placement. The principal objective is to minimize wire length, but this is not known exactly at the placement stage as the wiring paths have still to be decided. However, there are several estimation techniques which are commonly used. For example, the area used has proved to be a very simple but robust estimator. The neighbourhood structures will vary with the type of problem, but usually involve swapping two components or moving one in the horizontal or vertical direction. Several researchers report success with variants in which the distance over which a component can be moved in one go is reduced with temperature. In problems where there is freedom of placement, the solution space usually includes configurations in which components overlap and feasibility is controlled by a penalty factor for overlap in the cost function.

Once the components have been placed, a coarse model of the wiring problem is considered in which all positions and wiring paths are approximated by a grid. This is known as the global wiring problem, and annealing has been applied successfully in its solution using changes in the wiring path for a single connection at a time. Sechen *et al.* [28] have used annealing in their Timberwolf placement package which they report as out-performing other computerized solutions. The original version has since been improved and incorporated into Thunderbird, a complete standard cell layout package. They use a geometric cooling schedule in which the rate of cooling gets faster as the system cools. Other researchers have also reported success with annealing for VLSI problems and current research is directed at parallel implementations of the algorithm [29]. It should be noted that the solution time requirements for these implementations are very long and may involve several days computing. However, in this application area this does not appear to be a problem.

Other problems in computer design are also proving amenable to solution by annealing. An interesting example is a problem of locating services on a computer network studied by Vernekar *et al.* [30]. The problem can be modelled as an integer program involving both 0-1 and other integer variables. When the solution space includes all combinations of all feasible values for all of these variables a simulated annealing approach is very slow. However, if the 0-1 variables are fixed the problem can be solved as a transportation problem. Therefore the solution space was reduced to contain all feasible combinations of the 0-1 variables and the cost function was defined by the solutions to the transportation problems. This approach was compared to an exact solution method based on Lagrangean relaxation, and to a greedy heuristic. The annealing solutions were better than those produced by the greedy heuristic but solution times were increased by a factor of 2. The increase in solution times with problem size for the Lagrangean relaxation algorithm meant that it soon overtook annealing in this requirement, making it unsuitable for solving large network problems.

2.4.3 Sequencing and scheduling

Many of the applications reported in the Operational Research field involve scheduling or timetabling problems—particularly production scheduling. Many of these involve determining an optimum sequence for a given set of jobs through a set of machines, in order to minimize the total production or makespan time. In some instances the jobs must be kept together in 'families' to reduce set-up costs, but in others there is complete freedom of choice. The solution space is made up of the set of all feasible orderings and a variety of neighbourhoods can be defined, although those obtained by changing the position of a single job or swapping the positions of two jobs are the most popular. Where families of jobs are involved, two different neighbourhoods—one for family moves and one for job moves—must be defined, and the type of neighbourhood to use for a give iteration is decided at random with a pre-determined probability.

The straightforward flowshop sequencing problem provides a good illustration of the difficulties encountered in trying to determine the best parameters for a particular implementation. The problem has been tackled independently by Ogbu and Smith [17] and Osman and Potts [31]. Both use the same neighbourhood structure and cost function. Ogbu and Smith use a geometric cooling schedule and Osman and Potts a Lundy and Mees schedule. However, as has already been observed, the performance of the algorithm is unlikely to be affected by this as long as the ranges and rates of cooling are similar. The main difference is in the acceptance function. Osman and Potts use the standard exponential probability whereas Ogbu and Smith use a function based only on temperature and not involving the magnitude of the uphill moves. They also use a slightly different sampling mechanism. Both reports conclude that annealing provides better results than previous heuristics, but that this is at the expense of extra computational time. As the two studies were implemented in different languages on different machines a direct comparison was not possible on the basis of the two reports. Ogbu and Smith [32] therefore programmed the Osman and Potts algorithm and ran it on their own test data. They found that there was little difference in the quality of the solutions produced, but the time taken for the algorithms to converge did differ significantly. When the number of machines is small the Osman and Potts algorithm is faster than their own. However on data sets with 10 or more machines the reverse is true as the computation time of the Osman and Potts algorithm appears to increase far more rapidly with the number of machines. This illustrates the importance of carrying out any experiments on typical data, and suggests that it may be beneficial to use different parameters for problems of different sizes.

Annealing has also been used successfully for a number of other production scheduling problems. Kuik and Solomon [33] tackle a multi-level problem in which the parts manufactured may either be sold or used as components for a part at a different level. The objective is to schedule the manufacturing process into the available time periods in such a way as to minimize a combination of set up and inventory costs while meeting all demand. The problem can be modelled using a mixed integer programming formulation with 0-1 variables y_{it} indicating whether there was a set-up cost incurred for part i in period t. Kuik and Solomon use a strategy similar to that of Vernekar *et al.* [30]. Without these variables the problem is a straightforward LP and can easily be solved. The form of the LP means that its dual can be decomposed into separate problems for each part, and these can be solved iteratively for increasing time periods. In order to apply annealing to the problem, the set of feasible solutions is defined as the sets of values for the 0-1 variables, and neighbouring solutions are obtained by changing the value of one such variable. This has the effect of changing the coefficient of just one variable in the objective function of the LP, and the simple solution algorithm for the dual means that the effect of this change can be evaluated very simply. Annealing out-performs existing heuristic methods for this problem.

One of the most difficult problems in this class is that of scheduling a flexible manufacturing system which is concerned not only with jobs and machines but also the transportation system between machines. The problem is a complex one involving difficult feasibility constraints and costs based on efficiency calculations on the resulting network. One of the best heuristic approaches involves local optimization with a steepest descent strategy in which infeasibility is included as a penalty in the cost function. The neighbourhoods are defined by a series of heuristic rules which identify changes which are likely to result in improvements. Brandimarte *et al.* [16] found that simply converting the descent approach to an annealing scheme with the same neighbourhood structure was not successful, as the heuristic rules are so good that neighbourhoods tend not to contain any worse solutions. The rules were therefore relaxed and neighbourhoods were generated by moving jobs from busy periods into slack ones. Because no account is taken of the interaction of machines these rules do produce a reasonable number of uphill moves. Using these neighbourhoods annealing out-performs the original heuristic, in spite of the computational effort needed to generate the neighbours and to calculate the changes in the cost function.

Other areas of scheduling tackled using an annealing approach include transport scheduling and timetabling problems. Wright [34] describes a locomotive scheduling problem in which the cost function involves the solution of linear assignment problems. Unlike many simpler problems the change in cost is not calculated directly. A full calculation of the new cost is necessary. Rather than calculating $p(accept) = \exp\{(new\_cost - old\_cost)/t\}$ and then comparing this with a uniform random number x, Wright generates x and calculates $old\_cost - t \ln x$ and accepts if new_cost is no greater than this value. This means that the sub-totals in the summation of the cost function can be compared as they are calculated and the calculation aborted as soon as the value reaches that required for rejection. Domschke [35] investigates the problem of minimizing the waiting time for passengers who have to change trains in mid-journey on an urban rail network. The schedule is perturbed by moving the time of one train at a time. Annealing is compared with another heuristic and is found to give slightly better solutions in comparable time.

Timetable problems have also proved amenable to annealing. They have natural neighbourhoods obtained by swapping the allocation of two events in different time-slots, or by cancelling an event and replacing it—depending on the precise nature of the problem. The cost function measures the infeasibility of the current solution and may involve the weighted sum of penalties associated with different types of feasibility violation. The objective may be to produce a perfect time-table, i.e. to reduce the cost to zero, or to produce one which minimizes the penalties within a set of explicit constraints. Abramson [26] reports success in solving a variety of school timetabling problems in this way. For all but his largest test problem he managed to produce a perfect timetable. Eglese and Rand [36] and Dowsland [37] both consider problems in which it is known that a perfect solution is impossible, and the objective is to minimize a cost function which measures the disappointment of conference delegates in the first case and of university students in the second. Eglese and Rand report that their annealing implementation will be used to schedule the same conference in future years. Dowsland compares annealing with two exact algorithms and reports that fairly fast cooling and a geometric schedule produce optimal results with high probability. This is attributed to the coarseness of the cost function which results in several global optima spread over the solution space. However, a straightforward descent algorithm was not able to find these global optima with any degree of consistency and annealing definitely seems to be the better approach.

2.4.4 Other problems

A wide variety of other problems have been or are being tackled by annealing. There are a number of articles describing its use in image processing, for example the papers by Geman and Geman [38], and Carnevali *et al.* [39]. Sharpe *et al.* [40] are experimenting with the use of annealing in a commercial package for building layout design, and Derwent [2] suggests that it might provide for more efficient pollution control in Europe by optimally locating pollution abatement equipment. Currently such equipment is allocated uniformly over the region. Annealing was used to assess the effects of positioning the same amount of equipment differently, with the objective of minimizing the effect of nitrogen oxides over a series of measurement locations. Annealing illustrates that careful placement of the devices can be much more effective than the blanket policy currently adopted.

The usefulness of annealing is not limited to applications in OR and computer design and engineering. Discrete optimization problems arise in many disciplines and simulated annealing has already been applied to many of them including DNA mapping [41], problems of constructing evolutionary trees [42], and the design of statistical experiments [43]. A very comprehensive bibliography, classified by application area, has been produced by Collins *et al.* [44], while further applications in the field of Operational Research are cited by Eglese [45].

The evidence of these reports suggests that annealing is at its most useful in complex situations for which it is difficult to design robust heuristics which take account of the problem structure because of the many facets involved. This is the situation in VLSI design and with production scheduling problems. Computation times are long but the improvement in quality obtained over faster techniques is usually judged worthwhile. Annealing has also proved effective for problems in which infeasibility can be expressed in terms of a penalty function in the objective. This is the situation in many scheduling problems, and the observations of Eglese and Rand and of Dowsland suggest that a straightforward application of the basic annealing algorithm can give very good results for such problems without the need for extensive fine-tuning or very slow annealing schedules. On the other hand, well-structured problems often lend themselves to the design of problem-specific heuristics which take full advantage of the problem structure. In such cases annealing is unlikely to compete in terms of solution quality.

2.5 Conclusions

Simulated annealing as a heuristic technique for combinatorial optimization problems was heralded as a simple and robust algorithm which was capable of providing good solutions to some very difficult problems. The enormous interest in the technique from researchers in a variety of application areas was no doubt a result of these claims, together with the theoretical convergence results. However, the disappointingly long running times needed even to approximate convergence to the optimum, combined with the realization that fine-tuning of the cooling schedule and a judicious choice of neighbourhood structure were needed to get the best out of annealing, quenched some of this initial enthusiasm.

Since then a number of experiments and practical applications have shown that annealing can provide a useful solution method for a variety of problems, generally out-performing standard descent methods and sometimes competing effectively with specialist heuristics. In difficult situations where conflicting goals and problem variability make the design of problem-specific methods almost impossible it has proved very successful. Many of the reported successes involve modifications to the basic annealing algorithm. While these modifications often increase performance, they also increase the number of decisions which must be made by the algorithm designer. Thus the simplicity and robustness of the technique must be questioned.

In spite of this, it is still true to say that it is usually very easy to get a simple form of annealing working for almost any combinatorial optimization problem. The choices which must then be made in adapting the algorithm to a particular problem can be decided experimentally. Although the number of options available can be daunting, a detailed investigation into the capabilities of an annealing algorithm for a given problem can be fun, especially with imaginative use of graphics. It can also be very frustrating: one of the major problems is knowing when to stop, as there is always the feeling that the next modification or parameter change may lead to dramatic improvements in solution quality or computational requirements.

There is little doubt that the success of annealing has rekindled interest in local optimization heuristics. This has lead to a number of new techniques where the emphasis is on controlling uphill moves in an intelligent rather than purely random fashion. The best known such method is probably tabu search, discussed in the next chapter of this book, but other approaches such as simulated evolution have also been suggested. Some comparisons between the performances of tabu search and annealing have been attempted but, because both techniques define general approaches rather than a precise algorithm, it is difficult to draw any definite conclusions.

Research suggests that for a given, well-defined problem, it is likely that a well-designed problem-specific algorithm will outperform annealing. However, there is a trade off between generality and robustness on the one hand and solution quality on the other. With traditional algorithms the decision as to what degree of trade-off is desirable must be made at the outset. Simulated annealing allows this decision to be deferred until the project is well under way. At the one end of the scale an implementation of the basic algorithm with the most obvious neighbourhood structure, a geometric cooling rate of 0.95, and a starting temperature determined by a few quick experiments is very easy to implement, and will probably give reasonable results. At the other end an annealing algorithm with a cooling schedule determined as the result of extensive experimentation, a neighbourhood structure decided by in depth knowledge of the problem characteristics, combined with well researched modifications and appropriate heuristic rules and optimization steps can prove to be a very powerful problem specific approach.

When faced with the challenge of designing a heuristic solution for a new problem, simulated annealing is certainly worth considering. The basic claims of its early advocates still hold. It *is* easy to implement, it *is* applicable to almost any combinatorial optimization problem and it usually provides reasonable solutions. These can either be accepted immediately or the initial implementation can be used as a platform for further research and improvements.

References

[1] S.Kirkpatrick, C.D.Gellat and M.P.Vecchi (1983) Optimization by simulated annealing. *Science*, **220**, 671-680.

[2] D.Derwent (1988) A better way to control pollution. *Nature*, **331**, 575-578.

[3] N.Radcliffe and G.Wilson (1990) Natural solutions give their best. *New Scientist*, 14 April, 47-50.

[4] N.Metropolis, A.W.Rosenbluth, M.N.Rosenbluth, A.H.Teller and E.Teller (1953) Equation of state calculation by fast computing machines. *J.of Chem. Phys.*, **21**, 1087-1091.

[5] V.Černy (1985) A thermodynamical approach to the travelling salesman problem: an efficient simulation algorithm. *J.of Optimization Theory and Applic.*, **45**, 41-55.

[6] P.J.M.Van Laarhoven and E.H.L.Aarts (1988) *Simulated Annealing: Theory and Applications.* Kluwer, Dordrecht.

[7] E.H.L.Aarts and J.H.M.Korst (1989) *Simulated Annealing and Boltzmann machines.* Wiley, Chichester.

[8] E.H.L.Aarts and P.J.M.Van Laarhoven (1985) Statistical cooling: a general approach to combinatorial optimization problems. *Philips J. of Research*, **40**, 193-226

[9] B.Hajek (1988) Cooling schedules for optimal annealing. *MOR*, **13**, 311-329.

[10] M.Lundy and A.Mees (1986) Convergence of an annealing algorithm. *Math.Prog.*, **34**, 111-124.

[11] D.S.Johnson, C.R.Aragon, L.A.McGeoch and C.Schevon (1991) Optimization by simulated annealing: an experimental evaluation; part II, graph coloring and number partitioning. *Opns.Res.*, **39**, 378-406.

[12] M.Chams, A.Hertz and D.de Werra (1987) Some experiments with simulated annealing for colouring graphs. *EJOR*, **32**, 260-266.

[13] K.A.Dowsland (1992) Some experiments with simulated annealing techniques for packing problems. *EJOR*, (to appear).

[14] D.S.Johnson, C.R.Aragon, L.A.McGeoch and C.Schevon (1989) Optimization by simulated annealing: an experimental evaluation; part I, graph partitioning. *Opns.Res.*, **37**, 865-892.

[15] D.S.Johnson, C.R.Aragon, L.A.McGeoch and C.Schevon (1992) Optimization by simulated annealing: an experimental evaluation; part III, the travelling salesman problem. *Opns.Res.* (to appear).

[16] P.Brandimarte, R.Conterno and P.Laface (1987) FMS production scheduling by simulated annealing. *Proceedings of 3rd International Conference on Simulation in Manufacturing*, 235-245.

[17] F.A.Ogbu and D.K.Smith (1990) The application of the simulated annealing algorithm to the solution of the $n/m/Cmax$ flowshop problem. *Computers & Ops.Res.*, **17**, 243-253.

[18] A.J.Vakharia and Y-L.Chang (1990) A simulated annealing approach to scheduling a manufacturing cell. *NRL*, **37**, 559-577.

[19] D.T.Connolly (1990) An improved annealing scheme for the QAP. *EJOR*, **46**, 93-100.

[20] F.Glover and H.J.Greenberg (1989) New approaches for heuristic search: a bilateral link with artificial intelligence. *EJOR*, **39**, 119-130.

[21] C.Sechen, D.Braun and A.Sangiovanni-Vincetelli (1988) Thunderbird: A complete standard cell layout package. *IEEE J. of Solid-State Circuits*, **SC-23**, 410-420.

[22] C.A.Tovey (1988) Simulated simulated annealing. *AJMMS*, **8**, 389-407.

[23] Y.Rossier, M.Troyon and Th.M.Liebling (1986) Probabilistic exchange algorithms and euclidean travelling salesman problems. *OR Spektrum*, **8**, 151-164.

[24] J.W.Greene and K.J.Supowit (1986) Simulated annealing without rejected moves. *IEEE Trans. on Computer-Aided Design*, **CAD-5**, 221-228.

[25] E.Bonomi and J-L.Lutton (1984) The N-city travelling salesman problem: Statistical mechanics and the metropolis algorithm. *SIAM Review*, **26**, 551-568.

[26] D.Abramson (1991) Constructing school timetables using simulated annealing: sequential and parallel algorithms. *Man.Sci.*, **37**, 98-113.

[27] K.A.Dowsland (1991) Hill-climbing, simulated annealing, and the Steiner problem in graphs. *Eng.Opt.*, **17**, 91-107.

[28] C.Sechen and A.Sangiovanni-Vincetelli (1985) The TimberWolf placement and routing package. *IEEE J. of Solid-State Circuits*, **SC-20**, 510-522.

[29] S.A.Kravitz and R.A.Rutenbar (1987) Placement by simulated annealing on a multiprocessor. *IEEE Trans. on Computer Aided Design*, **CAD-6**, 534-549.

[30] A.Vernekar, G.Anandalingam and C.Dorny (1990) Optimization of resource location in hierarchical computer networks. *Computers & Ops.Res.*, **17**, 375-388

[31] I.H.Osman and C.N.Potts (1989) Simulated annealing for permutation flow-shop scheduling. *OMEGA*, **17**, 551-557.

[32] F.A.Ogbu and D.K.Smith (1991) Simulated annealing for permutation flow-shop problem. *OMEGA*, **19**, 65-67.

[33] R.Kuik and M.Salomon (1990) Multi-level lot-sizing problem: evaluation of a simulated annealing heuristic. *EJOR*, **45**, 25-37.

[34] M.B.Wright (1989) Applying stochastic algorithms to a locomotive scheduling problem. *JORS*, **40**, 187-192.

[35] W.Domschke (1989) Schedule synchronization for public transit networks. *OR Spektrum*, **11**, 17-24.

[36] R.W.Eglese and G.K.Rand (1987) Conference seminar timetabling. *JORS*, **38**, 591-598.

[37] K.A.Dowsland (1990) A timetabling problem in which clashes are inevitable. *JORS*, **41**, 907-918.

[38] S.Geman and D.Geman (1984) Stochastic relaxation, Gibbs distributions, and the Bayesian restoration of images. *IEEE Trans. on Pattern Analysis and Machine Intelligence*, **PAMI-6**, 721-741.

[39] P.Carnevali, L.Coletti and S.Patarnello (1985) Image processing by simulated annealing. *IBM J. of Research and Development*, **29**, 569-579.

[40] R.Sharpe, B.S.Marksjo, J.R.Mitchell and J.R.Crawford (1985) An interactive model for the layout of buildings. *Applied Mathematical Modelling*, **9**, 207-214.

[41] L.Goldstein and M.S.Waterman (1987) Mapping DNA by stochastic relaxation. *Adv.in Appl. Maths.*, **8**, 194-207.

[42] M.Lundy (1985) Applications of the annealing algorithm to combinatorial problems in statistics. *Biometrika*, **72**, 191-198.

[43] R.K.Meyer and C.J.Nachtsheim (1988) Constructing exact D-optimal experimental designs by simulated annealing. *AJMMS*, **8**, 329-359.

[44] N.E.Collins, R.W.Eglese and B.L.Golden (1988) Simulated annealing—an annotated bibliography. *AJMMS*, **8**, 209-307.

[45] R.W.Eglese (1990) Simulated annealing: a tool for operational research. *EJOR*, **46**, 271-281.

Chapter 3

Tabu Search

Fred Glover and Manuel Laguna

3.1 Introduction

Tabu search (TS) has its antecedents in methods designed to cross boundaries of feasibility or local optimality normally treated as barriers, and systematically to impose and release constraints to permit exploration of otherwise forbidden regions. Early examples of such procedures include heuristics based on surrogate constraint methods and cutting plane approaches that systematically violate feasibility conditions. The modern form of tabu search derives from Glover [1]. Seminal ideas of the method are also developed by Hansen [2] in a *steepest ascent/mildest descent* formulation. Additional contributions, such as those cited in the following pages, are shaping the evolution of the method and are responsible for its growing body of successful applications.

Webster's dictionary defines *tabu* or *taboo* as 'set apart as charged with a dangerous supernatural power and forbidden to profane use or contact...' or 'banned on grounds of morality or taste or as constituting a risk...'. Tabu search scarcely involves reference to supernatural or moral considerations, but instead is concerned with imposing restrictions to guide a search process to negotiate otherwise difficult regions. These restrictions operate in several forms, both by direct exclusion of certain search alternatives classed as 'forbidden', and also by translation into modified evaluations and probabilities of selection.

The purpose of this chapter is to integrate some of the fundamental ways of viewing and characterizing tabu search, with extended examples to clarify its operations. We also point to a variety of directions for new applications and research. Our development includes comparisons and contrasts between the principles of tabu search and those of simulated annealing (SA) and genetic algorithms (GAs). Computational implications of these differences, and foundations for creating hybrid methods that unite features of these different approaches are also discussed. In addition, we examine special designs and computational outcomes for incorporating tabu search as a driving mechanism within neural networks.

The philosophy of tabu search is to derive and exploit a collection of principles of intelligent problem solving. A fundamental element underlying tabu search is the use of flexible memory. From the standpoint of tabu search, flexible memory embodies the dual processes of creating and exploiting structures for taking advantage of history (hence combining the activities of acquiring and profiting from information).

The memory structures of tabu search operate by reference to four principal dimensions, consisting of recency, frequency, quality, and influence. These dimensions in turn are set against a background of logical structure and connectivity. The rôle of these elements in creating effective problem-solving processes provides the focus of our following development.

3.2 The Tabu Search Framework

To provide a background for understanding some of the fundamental elements of tabu search, we illustrate its basic operation with an example.

3.2.1 An illustrative example

Permutation problems form an important class of problems in optimization, and offer a useful vehicle to demonstrate some of the considerations that must be faced in the combinatorial domain. Classical instances of permutation problems include the travelling salesman problem, the quadratic assignment problem, production sequencing problems, and a variety of design problems. As a basis for illustration, consider the problem of designing a material consisting of a number of insulating modules. The order in which these modules are arranged determines the overall insulating property of the resulting material, as shown in Figure 3.1.

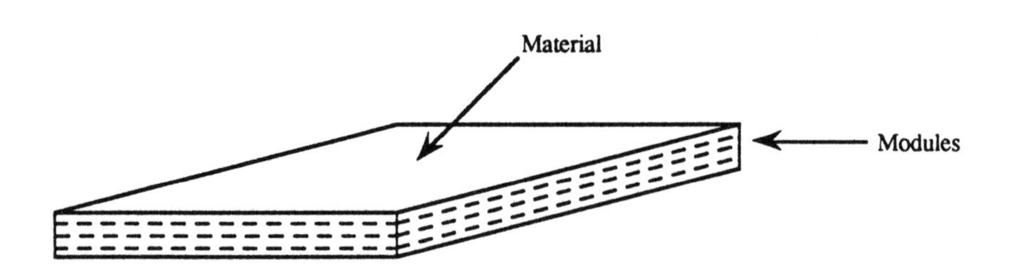

Figure 3.1: Modules in an insulating material

The problem is to find the ordering of modules that maximizes the overall insulating property of the composite material. Suppose that 7 modules are considered for a particular material, and that evaluating the overall insulating property of a particular ordering is a computationally expensive procedure. We desire a search method that is able to find an optimal or near-optimal solution by examining only a small subset of the total number of permutations possible (in this case 5040, though for many applications it can be astronomical).

Closely related problems that can be represented in essentially the same way include serial filtering and job sequencing problems. Serial filtering problems arise in pattern recognition and signal processing applications, where a given input is to be subjected to a succession of filters (or screening tests) to obtain the 'best' output. Filters are sequentially applied to the input signal, and the quality of the output is determined by the order in which they are placed (see Figure 3.2). In this case, the search method must be designed to find the best filtering sequence. Such filtering processes are also relevant to applications in chemical engineering, astronomy, and biochemistry.

Job sequencing problems consist of determining best sequences for processing a set of jobs on designated machines. Each machine is thus assigned some permutation of available jobs. (In some settings, multiple machine problems may be treated by extensions of processes

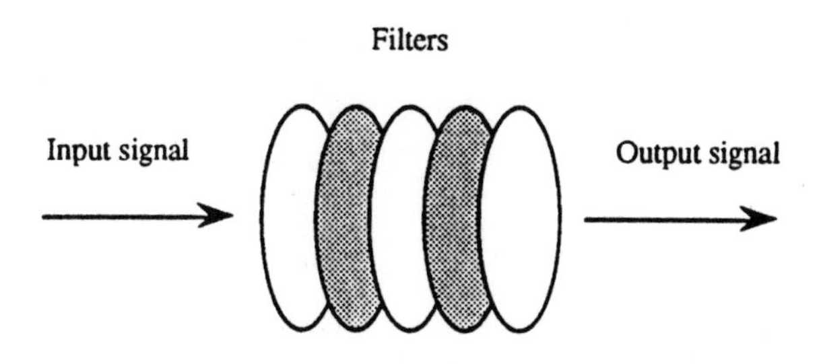

Figure 3.2: Filtering sequence

for single machine problems.) There are many variants of the single machine problem depending on the definition of 'best' sequence. For example, the best sequence may be the one that minimizes the *makespan*—the completion time of the last job in the sequence. Other possibilities are to minimize a weighted sum of tardiness penalties or a sum of setup costs.

For well-structured objective functions, evaluations of ways to move from one solution to another are generally fast. However, problems with even modest numbers of jobs overwhelm the capabilities of algorithms that 'guarantee' optimality, rendering them unable to obtain solutions in reasonable amounts of time. That is one of the reasons why effective heuristic approaches have proved important in the area of production scheduling.

Some useful variants of the foregoing problems can be represented 'as if' they were permutation problems. These include, for example, problems where it is simultaneously desired to select a best subset of items (modules, filters, jobs) from an available pool, and to identify a best sequence for this chosen set. In this case, the problem can be represented by creating a dummy position to hold a residual pool, where all items that do not currently occupy one of the sequence positions are placed. (The path assignment problem discussed in Section 3.4 is a good example of this kind of representation.)

We focus on the module insulation problem, using it to introduce and illustrate the basic components of tabu search. First we assume that an initial solution for this problem can be constructed in some intelligent fashion, i.e. by taking advantage of some problem-specific structure. Suppose the initial solution to our problem is the one shown in Figure 3.3.

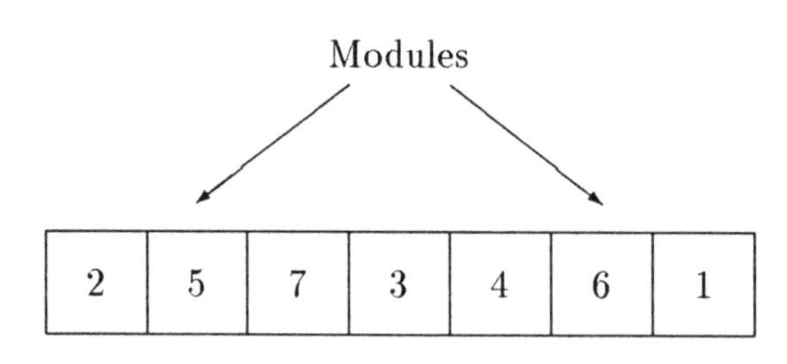

Figure 3.3: Initial permutation

The ordering in Figure 3.3 specifies that module 2 is placed in the first position, followed by module 5, etc. The resulting material has an insulating property of 10 units (which we assume was found by an accompanying evaluation routine, e.g. a simulator package for estimating the properties of a material without actually building a prototype). TS methods operate under the assumption that a *neighbourhood* can be constructed to identify 'adjacent solutions' that can be reached from any current solution. (Neighbourhood search is described in Section 3.2.3.) Pairwise exchanges (or swaps) are frequently used to define neighbourhoods in permutation problems, identifying *moves* that lead from one solution to the next. In our problem, a swap exchanges the position of two modules as illustrated in Figure 3.4. Therefore, the complete neighbourhood of a given current solution consists of the 21 adjacent solutions that can be obtained by such swaps.

Associated with each swap is a move value, which represents the change in the objective function value as a result of the proposed exchange. Move values generally provide a fundamental basis for evaluating the quality of a move, although other criteria can also be important, as indicated later. A chief mechanism for exploiting memory in tabu search is to classify a subset of the moves in a neighbourhood as forbidden (or tabu). The classification depends on the history of the search, particularly as manifested in the recency or frequency that certain move or solution components, called *attributes*,

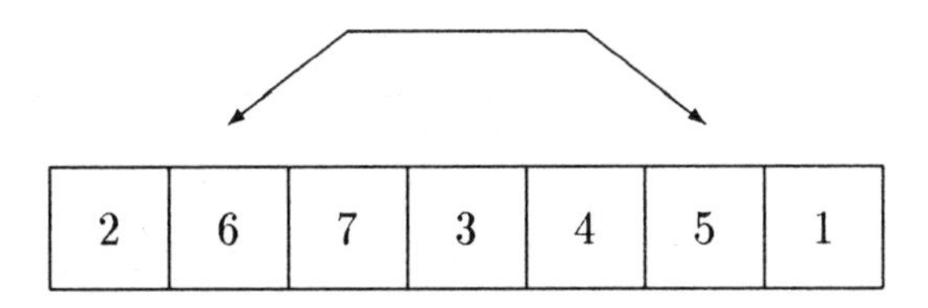

Figure 3.4: Swap of modules 5 and 6

have participated in generating past solutions. For example, one attribute of a swap is the identity of the pair of elements that change positions (in this case, the two modules exchanged). As a basis for preventing the search from repeating swap combinations tried in the recent past, potentially reversing the effects of previous moves by interchanges that might return to previous positions, we will classify as tabu all swaps composed of any of the most recent pairs of such modules; in this case, for illustrative purposes, the three most recent pairs. This means that a module pair will be kept tabu for a duration (tenure) of 3 iterations. Since exchanging modules 2 and 5 is the same as exchanging modules 5 and 2, both may be represented by the pair (2,5). Thus, a data structure such as the one shown in Figure 3.5 may be used.

Each cell of the structure in Figure 3.5 contains the number of iterations remaining until the corresponding modules are allowed to exchange positions again. Therefore, if the cell (3,5) has a value of zero, then modules 3 and 5 are free to exchange positions. On the other hand, if cell (2,4) has a value of 2, then modules 2 and 4 may not exchange positions for the next two iterations (i.e. a swap that exchanges these modules is classified tabu).

The type of move attributes illustrated here for defining tabu restrictions is not the only one possible. For example, reference may be made to separate modules rather than module pairs, or to positions of modules, or to links between their immediate predecessors (or successors), and so forth. Some choices of attributes are better than others, and relevant considerations are discussed in Section 3.2.5. (Attributes involving created and broken links between immediate predecessors and successors are often among the more effective for many permutation problems.)

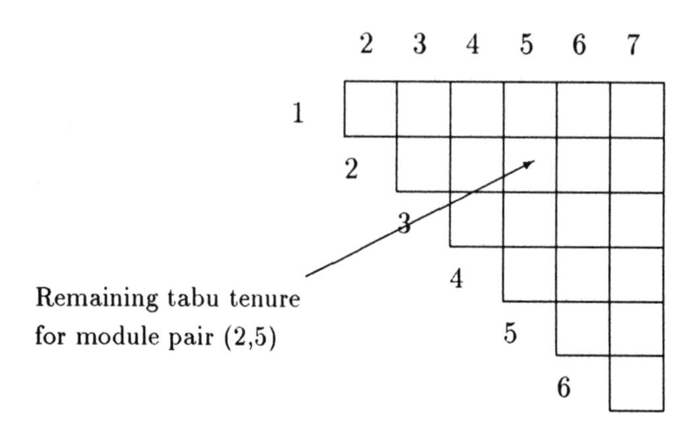

Figure 3.5: Tabu data structure for attributes consisting of module pairs exchanged

To implement tabu restrictions such as those based on module pairs, an important exception must be taken into account. Tabu restrictions are not inviolable under all circumstances. When a tabu move would result in a solution better than any visited so far, its tabu classification may be overridden. A condition that allows such an override to occur is called an *aspiration criterion.* (Several useful forms of such criteria are presented in Section 3.2.7.) The following shows 4 iterations of the basic tabu procedure that employs the *paired module* tabu restriction and the *best solution* aspiration criterion.

Iteration 0 (*Starting point*)

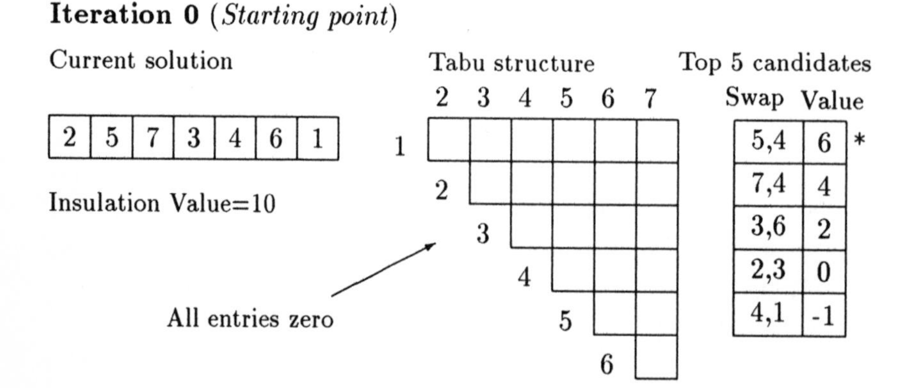

Swap	Value	
5,4	6	*
7,4	4	
3,6	2	
2,3	0	
4,1	-1	

The starting solution has an insulation value of 10, and the tabu data structure is initially empty, i.e. it is filled with zeros, indicating no moves are classified tabu at the beginning of the search. (For clarity, we have not actually inserted these zeros in this or the following diagrams.) After evaluating the candidate swap moves, the top five moves (in terms of move values) are shown in the table for iteration 0 above. This information is provided by an independent evaluation subroutine designed to identify move values for this particular problem. (Of course, it is not necessary for the subroutine to sort and identify each of the 5 best moves, since we are interested only in the best. The additional options are included here to clarify certain ideas subsequently presented.) To find a local maximum for the insulating property of the material, we swap the positions of modules 5 and 4 (as indicated by the asterisk). The total gain of such a move equals 6 units.

Iteration 1

Current solution

2	4	7	3	5	6	1

Insulation Value=16

Tabu structure

	2	3	4	5	6	7
1						
2						
3						
4				3		
5						
6						

Top 5 candidates

Swap	Value	
3,1	2	*
2,3	1	
3,6	-1	
7,1	-2	
6,1	-4	

The new current solution has an insulating value of 16 (i.e. the previous insulation value plus the value of the selected move). The tabu structure now shows that swapping the positions of modules 4 and 5 is forbidden for 3 iterations. The most improving move at this step is to swap 3 and 1 for a gain of 2.

Iteration 2

Current solution

2	4	7	1	5	6	3

Insulation Value=18

Tabu structure

	2	3	4	5	6	7
1		3				
2						
3						
4				2		
5						
6						

Top 5 candidates

Swap	Value	
1,3	-2	T
2,4	-4	*
7,6	-6	
4,5	-7	T
5,3	-9	

The new current solution becomes the best solution found so far with an insulating value of 18. At this iteration, two exchanges are classified tabu, as indicated by the nonzero entries in the tabu structure.

Note that entry (4,5) has been decreased from 3 to 2, indicating that its original tabu tenure of 3 now has 2 remaining iterations to go. This time, none of the candidates (including the top 5 shown) has a positive move value. Therefore, a non-improving move has to be made. The most attractive non-improving move is the reversal of the move performed in the previous iteration, but since it is classified tabu, this move is not selected. Instead, the swap of modules 2 and 4 is chosen, as indicated by the asterisk.

Iteration 3

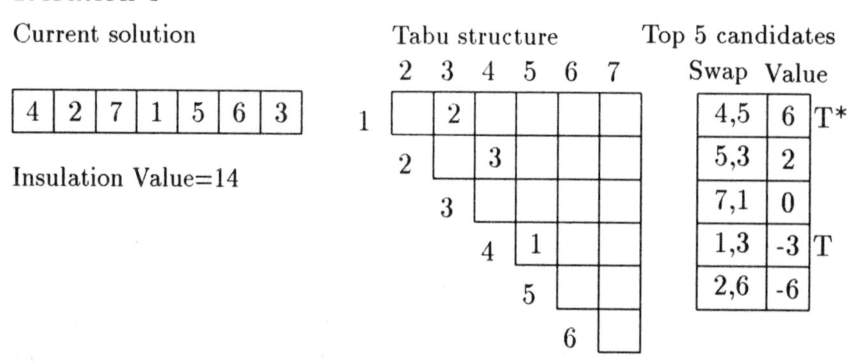

Current solution

4	2	7	1	5	6	3

Insulation Value=14

Tabu structure

	2	3	4	5	6	7
1		2				
2			3			
3						
4				1		
5						
6						

Top 5 candidates

Swap	Value	
4,5	6	T*
5,3	2	
7,1	0	
1,3	-3	T
2,6	-6	

The new current solution has an insulation value inferior to the two values previously obtained, as a result of executing a move with a negative move value. The tabu data structure now indicates that 3 moves are classified tabu, with different remaining tabu tenures. At the top of the candidate list, we find the swap of modules 4 and 5, which in effect represents the reversal of the first move performed, and is classified tabu. However, performing this move produces a solution with an objective function value that is superior to any previous insulation value. Therefore, we make use of the aspiration criterion to override the tabu classification of this move and select it as the best on this iteration.

Iteration 4

Current solution

5	2	7	1	4	6	3

Insulation Value=20

Tabu structure

	2	3	4	5	6	7
1		1				
2			2			
3						
4				3		
5						
6						

Top 5 candidates

Swap	Value	
7,1	0	*
4,3	-3	
6,3	-5	
5,4	-6	T
2,6	-8	

The current solution becomes the incumbent new best solution and the process continues. Note that the chosen tabu restriction and tabu tenure of 3 results in forbidding only 3 out of 21 possible swaps, since the module pair with a residual tenure of 1 always drops to a residual tenure of 0 each time a new pair with tenure 3 is introduced. (By recording the iteration when a module pair becomes tabu, and comparing this against the current iteration to determine the remaining tabu tenure, it is unnecessary to change these entries at each step as we do here.)

In some situations, it may be desirable to increase the percentage of available moves that receive a tabu classification. This may be achieved either by increasing the tabu tenure or by changing the tabu restriction. For example, a tabu restriction that forbids swaps containing at least one member of a module pair will prevent a larger number of moves from being executed, even if the tenure remains the same. (In our case, this restriction would forbid 15 out of 21 swaps if the tabu tenure remains at 3.) Such a restriction is based on single module attributes instead of paired module attributes, and can be implemented with much less memory, i.e. by an array that records a tabu tenure for each module separately. Generally speaking, regardless of the type of restriction selected, improved outcomes are often obtained by tabu tenures that vary dynamically, as described in Section 3.2.6.

Move Values and Updates Because tabu search aggressively selects best admissible moves (where the meaning of best is affected by tabu classification and other elements to be indicated), it must examine and compare a number of move options. For many problems, only a portion of the move values will change from one iteration to the next, and often these changed values can be isolated and updated very quickly. For example, in the present illustration it may be useful to store a table *move_value*(j, k), which records the current move value for exchanging modules j and k. When a move is executed, a relatively small part of this table (consisting of values that change) can be quickly modified, and the updated table can then be consulted to identify moves that become the new top candidates.

Such partial updating often can be further enhanced by a list *move_name*(*move_value*) which, for each *move_value* in a relevant range, identifies *move_name* to be a specific move that yields this value. A linked list then can connect this *move_name* to the names of all other moves that yield the same *move_value*. The combination of the *move_name*(*move_value*) array and the linked list can be updated very quickly to make it easy to locate moves with best move values in cases where only a relatively small number of elements change. A given *move_value* entry can also refer to a range of move values, with an option to regard all values within a specified range as 'essentially equivalent'. (However, we suggest the merit of differentiating members of a given range more carefully upon approaching local optimality.)

On a broader scale, lists to facilitate access to best moves invite differentiation to include considerations introduced by move influence (Section 3.2.7) and by candidate list strategies (Section 3.3). They also are subject to periodic scanning with reference to concerns that extend beyond the short term horizon, as we illustrate next.

Complementary Tabu Memory Structures The accompaniment of recency-based memory with frequency-based memory adds a component that typically operates over a longer horizon. To illustrate one of the useful longer term applications of frequency-based memory, suppose that 25 TS iterations have been performed, and that the number of times each module pair has been exchanged is saved in an expanded tabu data structure. The lower diagonal of this structure now contains the frequency counts.

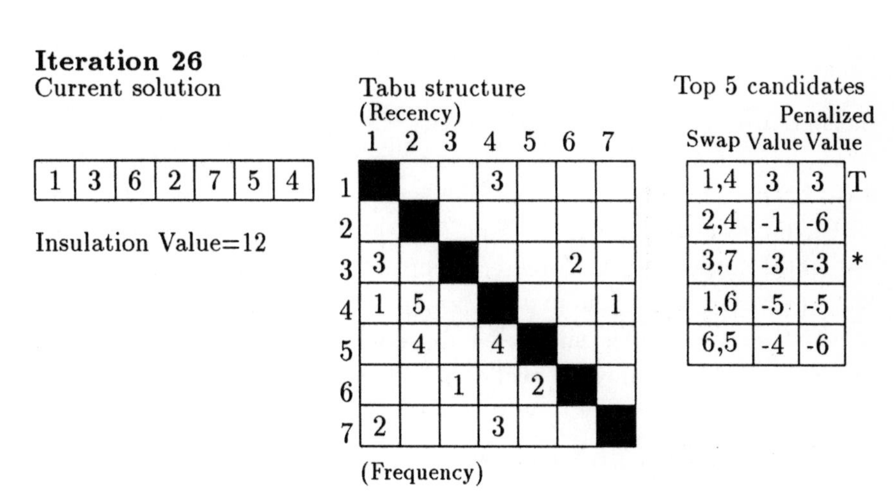

At the current iteration (iteration 26), the recency memory indicates that the last three module pairs exchanged were (1,4), (3,6), and (4,7). The frequency counts show the distribution of moves throughout the first 25 iterations. We use these counts to *diversify* the search, driving it into new regions. This diversifying influence is restricted to operate only on particular occasions. In this case, we select those occasions where no admissible improving moves exist. Our use of the frequency information will penalize non-improving moves by assigning a larger penalty to swaps of module pairs with greater frequency counts. (Typically these counts would be normalized, as by dividing by the total number of iterations or their maximum values.) We illustrate this in the present example by simply subtracting a frequency count from the associated move value.

The list of top candidates for iteration 26 shows that the most improving move is the swap (1,4), but since this module pair has a residual tabu tenure of 3, it is classified tabu. The move (2,4) has a value of -1, and it might otherwise be the one next preferred, except that its associated modules have been exchanged frequently during the history of the search (in fact, more frequently than any other module pair). Therefore, the move is heavily penalized and it loses its attractiveness. The swap of modules 3 and 7 is thus selected as the best move on the current iteration.

The strategy of instituting penalties only under particular conditions is used to preserve the aggressiveness of the search. Penalty functions in general are designed to account not only for frequencies but also for move values and certain influence measures, as discussed in Section 3.2.8.

In addition, frequencies defined over different subsets of past solutions, particularly subsets of élite solutions consisting of high quality local optima, give rise to complementary strategies of *intensification*. Intensification and diversification strategies interact to provide fundamental cornerstones of longer term memory in tabu search. The ways in which such elements are capable of creating enhanced search methods, extending the simplified approach of the preceding example, are elaborated in following sections.

3.2.2 Notation and problem description

A few basic definitions and conventions are useful as a foundation for communicating the principal ideas of TS. For this purpose we express the mathematical optimization problem in a slightly more general form than that used in chapter 1.

$$\begin{array}{ll} \textit{Minimise} & c(x) \\ \textit{subject to} & x \in \mathbf{X} \end{array}$$

The objective function $c(x)$ may be linear or nonlinear, and the condition $x \in \mathbf{X}$ summarizes constraints on the vector x. These constraints may include linear or nonlinear inequalities (as in chapter 1), and may compel some or all components of x to receive discrete values.

In many applications of combinatorial optimization, the problem of interest is not explicitly formulated as we have shown it. In such cases the present formulation may be conceived as a code for another formulation. The requirement $x \in \mathbf{X}$, for example, may specify logical conditions or interconnections that would be cumbersome to formulate mathematically, but may better be left as verbal stipulations (for example, in the form of rules). Often in these instances the variables are simply codes for conditions or assignments that are parts of the more complex structure. For example, an element of x may be a binary variable that receives a value of 1 to code for assigning an element u to a set or position v, and that receives a value 0 to indicate the assignment does not occur.

3.2.3 Neighbourhood search

TS may be conveniently characterized as a form of neighbourhood search, which has already been described in chapter 2. However, here we wish to define neighbourhood search in a less restricted fashion than usual. Frequently, for example, constructive and destructive procedures are excluded, whereas such procedures and their combinations are routinely subjected to the guidance of TS.

In neighbourhood search, each solution $x \in \mathbf{X}$ has an associated set of neighbours, $N(x) \subset \mathbf{X}$, called the neighbourhood of x. Each solution $x' \in N(x)$ can be reached directly from x by an operation called a *move*, and x is said to move (or transition) to x' when such an operation is performed. Normally in TS, neighbourhoods are assumed symmetric, i.e. x' is a neighbour of x if and only if x is a neighbour of x'.

Neighbourhood Search Method

Step 1	(Initialization)
(A)	Select a starting solution $x^{now} \in \mathbf{X}$.
(B)	Record the current best known solution by setting $x^{best} = x^{now}$ and define $best\_cost = c(x^{best})$.
Step 2	(Choice and termination)
	Choose a solution $x^{next} \in N(x^{now})$. If the choice criteria employed cannot be satisfied by any member of $N(x^{now})$ (hence no solution qualifies to be x^{next}), or if other termination criteria apply (such as a limit on the total number of iterations), then the method stops.
Step 3	(Update)
	Re-set $x^{now} = x^{next}$, and if $c(x^{now}) < best\_cost$, perform Step 1(B). Then return to Step 2.

The steps of neighbourhood search are as described above, where we assume choice criteria for selecting moves, and termination criteria for ending the search, are given by some external set of prescriptions.

The foregoing procedure can represent a constructive method by stipulating that **X** is expanded to include x vectors whose components take null (unassigned) values, and by stipulating that a neighbour x' of x can result by replacing a null component of x with a non-null component. (A change of representation sometimes conveniently allows null components to be represented by values of 0 and non-null components by values of 1.) A standard constructive method does not yield symmetric neighbourhoods, since non-null components are not permitted to become null again (hence the method ends when no more components are null). However, tabu search reinstates the symmetric relation by allowing constructive and destructive moves to co-exist, as a special instance of an approach called strategic oscillation (see Section 3.3).

The neighbourhood search method can easily be altered by adding special provisions to yield a variety of classical procedures. Descent methods, which only permit moves to neighbour solutions that improve the current $c(x^{now})$ value, and which end when no improving solutions can be found, can be expressed by the following provision in Step 2.

Descent Method

Step 2 (Choice and termination)
Choose $x^{next} \in N(x^{now})$ to satisfy $c(x^{next}) < c(x^{now})$ and terminate if no such x^{next} can be found.

The evident shortcoming of a descent method is that the final x^{now} obtained is a local optimum, which in most cases will not be a global optimum.

Randomized procedures such as Monte Carlo methods, which include simulated annealing, can similarly be represented by adding a simple provision to Step 2.

Monte Carlo Method

Step 2	(Choice and termination)
(A)	Randomly select x^{next} from $N(x^{now})$.
(B)	If $c(x^{next}) \le c(x^{now})$ accept x^{next} (and proceed to Step 3).
(C)	If $c(x^{next}) > c(x^{now})$ accept x^{next} with a probability that decreases with increases in the difference $c(x^{next}) - c(x^{now})$. If x^{next} is not accepted on the current trial by this criterion, return to (A).
(D)	Terminate by a chosen cutoff rule.

Monte Carlo methods continue to sample the search space until finally terminating by some form of iteration limit. Normally they use an exponential function to define probabilities, drawing from practice established in engineering and physical science. As described in chapter 2, the Monte Carlo version represented by simulated annealing starts with a high probability for accepting non-improving moves in Step 2(C) which is decreased over time as a function of a parameter called the 'temperature' which monotonically diminishes toward zero as the number of iterations grows.

Such approaches offer a chance to do better than finding a single local optimum since they effectively terminate only when the probability of accepting a non-improving move in Step 2(C) becomes so small that no such move is ever accepted (in the finite time allowed). Hence, they may wander in and out of various intermediate local optima prior to becoming lodged in a final local optimum, when the temperature becomes small.

Another randomizing approach to overcome the limitation of the descent method is simply to restart the method with different randomly selected initial solutions, and run the method multiple times. Such a random restart approach, sometimes called *iterated descent*, may be contrasted with a random perturbation approach, which simply chooses moves randomly for a period after reaching each local optimum, and then resumes a trajectory of descent. Alternating threshold methods indicated in Section 3.2.7 provide a refinement of this idea.

3.2.4 Tabu search characteristics

Tabu search, in contrast to the preceding methods, employs a somewhat different philosophy for going beyond the criterion of terminating at a local optimum. Randomization is de-emphasized, and generally is employed only in a highly constrained way, on the assumption that intelligent search should be based on more systematic forms of guidance. Thus randomization (pseudo-randomization) is chiefly assigned the rôle of facilitating operations that are otherwise cumbersome to implement or whose strategic implications are unclear. (In the latter case, a supplementary learning approach such as target analysis—see Laguna and Glover [3]—is customarily employed to determine if such implications can be sharpened.) Accordingly, many tabu search implementations are largely or wholly deterministic. An exception occurs for the variant called probabilistic tabu search, which selects moves according to probabilities based on the status and evaluations assigned to these moves by the basic tabu search principles. (A discussion of probabilistic convergence issues is provided by Faigle and Kern [4].)

Special TS uses of memory: modifying neighbourhood structures

The notion of exploiting certain forms of flexible memory to control the search process is the central theme underlying tabu search. The effect of such memory may be envisioned by stipulating that TS maintains a selective history H of the states encountered during the search, and replaces $N(x^{now})$ by a modified neighbourhood which may be denoted $N(H, x^{now})$. History therefore determines which solutions may be reached by a move from the current solution, selecting x^{next} from $N(H, x^{now})$.

In TS strategies based on short term considerations, $N(H, x^{now})$ is typically a subset of $N(x^{now})$, and the tabu classification serves to identify elements of $N(x^{now})$ excluded from $N(H, x^{now})$. In the intermediate and longer term strategies, $N(H, x^{now})$ may contain solutions not in $N(x^{now})$, generally consisting of selected élite solutions (high quality local optima) encountered at various points in the solution process. Such élite solutions are typically identified as elements of a regional cluster in intermediate term intensification strategies, and as elements of different clusters in longer term diversification strategies. In addition, élite solution components, in contrast to the solutions themselves, are included among the elements that can be retained and integrated to provide inputs to the search process.

TS also uses history to create a modified evaluation of currently accessible solutions. This may be expressed formally by saying that TS replaces the objective function $c(x)$ by a function $c(H,x)$, which has the purpose of evaluating the relative quality of currently accessible solutions. (An illustration is provided by the use of frequency-based memory in the example of Section 3.2.1.) The relevance of this modified function occurs because TS uses aggressive choice criteria that seek a best x^{next}, i.e. one that yields a best value of $c(H,x^{next})$, over a candidate set drawn from $N(H,x^{now})$. Moreover, modified evaluations are often accompanied by systematic alteration of $N(H,x^{now})$, to include neighbouring solutions that do not satisfy customary feasibility conditions (i.e. that strictly speaking do not yield $x \in \mathbf{X}$). Reference to $c(x)$ is retained for determining whether a move is improving or leads to a new best solution.

For large problems, where $N(H,x^{now})$ may have many elements, or for problems where these elements may be costly to examine, the aggressive choice orientation of TS makes it highly important to isolate a candidate subset of the neighbourhood, and to examine this subset instead of the entire neighbourhood. This can be done in stages, allowing the candidate subset to be expanded if alternatives satisfying aspiration levels are not found. Because of the significance of the candidate subset's rôle, we refer to this subset explicitly by the notation *Candidate_N*(x^{now}). Then the tabu search procedure may be expressed in the following manner.

Tabu Search Method

Step 1 (Initialization)
Begin with the same initialization used by Neighbourhood Search, and with the history record H empty.

Step 2 (Choice and termination)
Determine *Candidate_N*(x^{now}) as a subset of $N(H,x^{now})$. Select x^{next} from *Candidate_N*(x^{now}) to minimize $c(H,x)$ over this set. (x^{next} is called a highest evaluation element of *Candidate_N*(x^{now}).) Terminate by a chosen iteration cut-off rule.

Step 3 (Update)
Perform the update for the Neighbourhood Search Method, and additionally update the history record H.

Formally the tabu search method is quite straightforward to state. The essence of the method depends on how the history record H is defined and used, and on how the neighbourhood $Candidate\_N(x^{now})$ and the evaluation function $c(H,x)$ are determined.

In the simplest cases we may imagine $Candidate\_N(x^{now})$ to constitute all of $N(H,x^{now})$, and take $c(H,x)=c(x)$, disregarding neighbourhood screening approaches and the longer term considerations that introduce élite solutions into the determination of moves. We begin from this point of view, focusing on the short term component of tabu search for determining the form and use of H. The basic considerations provide a foundation for the intermediate and long term TS components as well.

3.2.5 Tabu search memory

Attribute based memory

An attribute of a move from x^{now} to x^{next}, or more generally of a trial move from x^{now} to a tentative solution x^{trial}, can encompass any aspect that changes as a result of the move. Natural types of attributes are as follows.

Illustrative Move Attributes for a Move x^{now} to x^{trial}

(A1)	Change of a selected variable x_j from 0 to 1.
(A2)	Change of a selected variable x_k from 1 to 0.
(A3)	The combined change of (A1) and (A2) taken together.
(A4)	Change of $c(x^{now})$ to $c(x^{trial})$.
(A5)	Change of a function $g(x^{now})$ to $g(x^{trial})$ (where g may represent a function that occurs naturally in the problem formulation or that is created strategically).
(A6)	Change represented by the difference value $g(x^{trial})-g(x^{now})$.
(A7)	The combined changes of (A5) or (A6) for more than one function g considered simultaneously.

A single move can evidently give rise to multiple attributes. For example, a move that changes the values of two variables simultaneously may give rise to each of the three attributes (A1), (A2), and (A3), as well as to other attributes of the form indicated. Attributes that represent combinations of other attributes do not necessarily provide more exploitable information, as will be seen. Attributes (A5) to (A7) are based on a function g that may be strategically chosen to be completely independent from c. For example, g may be a measure of distance (or dissimilarity) between any given solution and a reference solution, such as the last local optimum visited or the best solution found so far. Then, attribute (A6) would indicate whether a trial solution leads the search further from or closer to the reference point.

Move attributes, involving change, may be subdivided into component attributes called *from-attributes* and *to-attributes*. That is, each move attribute may be expressed as an ordered pair (*from-attribute, to-attribute*) whose components are respectively attributes of the solutions x^{now} and x^{trial}. Letting $A(x^{now})$ and $A(x^{trial})$ denote attribute sets for these two solutions, the requirement of change underlying the definition of a move attribute implies

$$\begin{aligned} \textit{from-attribute} &\in A(x^{now}) - A(x^{trial}) \\ \textit{to-attribute} &\in A(x^{trial}) - A(x^{now}). \end{aligned}$$

This differentiation between move attributes and their component *from-attributes* and *to-attributes* is useful for establishing certain outcomes related to their use.

When we refer to assigning alternative values to a selected variable x_j of x, and particularly to assigning values 0 and 1 to a binary variable, we will understand by our previous conventions that this can refer to a variety of operations such as adding or deleting edges from a graph, assigning or removing a facility from a particular location, changing the processing position of a job on a machine, and so forth. Such coding conventions can be extended to include the creation of supplementary variables that represent states of subservient processes. For example, $x_j = 0$ or 1 may indicate that an associated variable is nonbasic or basic in an extreme point solution procedure, as in the simplex method and its variants for linear and nonlinear programming.

Uses of move attributes

Recorded move attributes are often used in tabu search to impose constraints, called tabu restrictions, that prevent moves from being chosen that would reverse the changes represented by these attributes. More precisely, when a move from x^{now} to x^{next} is performed that contains an attribute e, a record is maintained for the reverse attribute which we denote by $\bar{e}$, in order to prevent a move from occurring that contains some subset of such reverse attributes. Examples of kinds of tabu restrictions frequently employed are as follows.

Illustrative Tabu Restrictions

A move is tabu if:

(R1) x_j changes from 1 to 0 (where x_j previously changed from 0 to 1).

(R2) x_k changes from 0 to 1 (where x_k previously changed from 1 to 0).

(R3) at least one of (R1) and (R2) occur. (This condition is more restrictive than either (R1) or (R2) separately—i.e. it makes more moves tabu.)

(R4) both (R1) and (R2) occur. (This condition is less restrictive than either (R1) or (R2) separately—i.e. it makes fewer moves tabu.)

(R5) both (R1) and (R2) occur, and in addition the reverse of these moves occurred simultaneously on the same iteration in the past. (This condition is less restrictive than (R4).)

(R6) $g(x)$ receives a value v' that it received on a previous iteration (i.e. $v' = g(x')$ for some previously visited solution x').

(R7) $g(x)$ changes from v'' to v', where $g(x)$ changed from v' to v'' on a previous iteration (i.e. $v' = g(x')$ and $v'' = g(x'')$ for some pair of solutions x' and x'' previously visited in sequence.)

Among the restrictions of these examples, only (R5) applies to a composite attribute, in which two component attributes simultaneously identify a single attribute of a previous move. (However, (R4) is meaningful only if the present move is composed of two such attributes, but does not depend on the condition that both of these attributes have occurred together in the past.) Also, while (R7) is less restrictive than (R6) (since it renders fewer moves tabu), both of these restrictions can reduce either to (R1) or (R2) by specifying $g(x) = x_j$ or $g(x) = x_k$. (Restriction (R6) is equivalent to (R7) in the situation where $g(x)$ can only take two different values.)

Tabu restrictions are also sometimes used to prevent repetitions rather than reversals, as illustrated by stipulating in (R1) that x_j previously changed from 1 to 0, rather than from 0 to 1. These have a rôle of preventing the repetition of a search path that leads away from a given solution. By contrast, restrictions that prevent reversals have a rôle of preventing a return to a previous solution. Hence, tabu restrictions vary according to whether they are defined in terms of reversals or duplications of their associated attributes.

The rôle of tabu status

A tabu restriction is typically activated only in the case where its attributes occurred within a limited number of iterations prior to the present iteration (creating a recency-based restriction), or occurred with a certain frequency over a longer span of iterations (creating a frequency-based restriction). More precisely, a tabu restriction is enforced only when the attributes underlying its definition satisfy certain thresholds of recency or frequency. To exploit this notion, we define an attribute to be tabu-active when its associated reverse (or duplicate) attribute has occurred within a stipulated interval of recency or frequency in past moves. An attribute that is not tabu-active is called tabu-inactive.

The condition of being tabu-active or tabu-inactive is called the *tabu status* of an attribute. Sometimes an attribute is called tabu or not tabu to indicate that it is tabu-active or tabu-inactive. It is important to keep in mind in such cases that a 'tabu attribute' does not correspond to a tabu move. As the preceding examples show, a move may contain tabu-active attributes, but still may not be tabu if these attributes are not of the right number or kind to activate a tabu restriction.

The most common tabu restrictions, whose attributes are the reverse of those defining these restrictions, characteristically have a goal of preventing cycling and of inducing vigour into the search. However, some types of restrictions must be accompanied by others, at least periodically, to achieve the cycle avoidance effect. For example, the restriction (R5) is not able to prevent cycling by itself, regardless of the interval of time it is allowed to be in effect. This can be demonstrated by letting the ordered pair (j, k) denote an attribute in which x_j changes from 0 to 1 and x_k changes from 1 to 0. Then a sequence of 3 moves that creates the three attributes (1,2), (2,3), and (3,1) both starts and ends at the same solution, but this sequence is not prevented by restriction (R5). (R7) also may not prevent cycling, if $g(x)$ can change from a later value to an earlier value without visiting values that were successively generated at intermediate points (e.g. going from 5 to 10 to 15 and then back to 5, jumping over the reverse move from 15 to 10).

Cycle avoidance can easily be achieved over the duration of tabu tenure, however, by focusing specifically on *from-attributes* and *to-attributes* rather than on their ordered pair combinations. More precisely, as long as at least one *to-attribute* of a current move is not a *from-attribute* of a previous move, cycling cannot occur. Examination of the preceding restrictions shows that all except (R5) and (R7) implicitly are based on the requirement that specified *from-attributes* of previous moves must not be *to-attributes* of the current move, or else the move is tabu. (The only component attributes of the present move that are relevant to its tabu classification are its *to-attributes*, which to prevent reversals must be *from-attributes* of previous moves.)

It should be pointed out, however, that cycle avoidance is not an ultimate goal of the search process. In some instances, a good search path will result in revisiting a solution encountered before. The broader objective is to continue to stimulate the discovery of new high quality solutions. Hence in the longer term the issue of cycle avoidance is more subtle than simply preventing a solution from being revisited. The way that tabu restrictions depend on different choices of move attributes, and the consequences of this dependency, are examined in the following example.

An Example Consider a past move that involves a change from $x_j = p$ to $x_j = q$. To avoid a reversal, we stipulate that the *from-attribute* of this move, $x_j = p$, is tabu-active, thus allowing the possibility of preventing a move with a change in which $x_j = p$ is the *to-attribute*. But $x_j = p$ is not the only component of the past move that can qualify as a *from-attribute*, and hence that can be the basis for defining a tabu-active status.

By conceiving an attribute change implicitly to involve replacing an attribute e by a complementary attribute $\overline{e}$, the change from $x_j = p$ to $x_j = q$ in fact may be viewed as composed of two such attribute changes: from $x_j = p$ to $x_j \neq p$, and from $x_j \neq q$ to $x_j = q$. Thus, $x_j \neq q$ can also be regarded as a *from-attribute* of this change. By avoiding either of the tabu-active reverse attributes, to $x_j = p$ or to $x_j \neq q$, the present move will not be able to re-visit the solution that initiated the past move. (Note that avoiding $x_j \neq q$ is the same as compelling $x_j = q$, which is more restrictive than avoiding $x_j = p$.)

The problem illustrated at the start of this chapter gives an instructive example of options created by identifying tabu attributes in this way. The swap moves of the illustration consist of selecting two items, j and k, where items j and k occupy positions p and q respectively, and then exchanging their positions. Let $x_u = v$ denote the statement 'item u is assigned to position v'. Hence the the swap move for interchanging the positions of items j and k can be represented as consisting of the two operations 'from $x_j = p$ to $x_j = q$' and 'from $x_k = q$ to $x_k = p$'. Subdividing these operations into their components, we can express the outcome as consisting of the following changes:

$$\text{from } x_j = p \text{ to } x_j \neq p$$
$$\text{from } x_j \neq q \text{ to } x_j = q$$
$$\text{from } x_k = q \text{ to } x_k \neq q$$
$$\text{from } x_k \neq p \text{ to } x_k = p.$$

Thus, any combination of the preceding *from-attributes* can be selected to represent corresponding *to-attributes* of a move currently under consideration, for the purpose of defining a tabu restriction applicable to this move. We may elect, for instance, to rely on just the first and third of the preceding *from-attributes*, using the tabu restriction that classifies a move tabu only if it contains both $x_j = p$ and $x_k = q$ as *to-attributes.* (Hence this prevents the current move if it transfers item j to position p and item k to position q, where items j and k were respectively moved out of these two positions in the past, though not necessarily on the same move.) This is a weaker restriction than one based on either the second or fourth *from-attribute* above, which renders a move tabu if it contains $x_j \neq q$ or $x_k \neq p$ as a *to-attribute*, hence essentially compelling the current move to result in $x_j = q$ or $x_k = p$ (or possibly both, depending on the restriction chosen). One implication of choosing stronger or weaker tabu restrictions is to render smaller or larger tabu tenures appropriate.

Effect of Variable Codings Different codings of variables also lead to different consequences for creating tabu restrictions. For example, if $x_u = v$ instead is given the interpretation 'item u immediately precedes item v', then the swap of items j and k yields an altered set of attributes with different associated possibilities. Denoting the two items that immediately precede and immediately follow j by p and q, and the corresponding items for k by r and s, we see that the swap creates the following changes:

$$\text{from } x_j = q \text{ to } x_j = s$$

$$\text{from } x_k = s \text{ to } x_k = q$$
$$\text{from } x_p = j \text{ to } x_p = k$$
$$\text{from } x_r = k \text{ to } x_r = j.$$

Moreover, each of these subdivides into two additional components (for example, the first becomes 'from $x_j = q$ to $x_j \neq q$' and 'from $x_j \neq s$ to $x_j = s$'), yielding a set of options for defining tabu restrictions that is considerably expanded over those of the preceding coding of the variables.

Representationally, there may be multiple options for characterizing the same set of attributes, and it is appropriate to use one that is natural for the problem setting. In this case, for example, it is convenient to represent the condition 'item u immediately precedes item v' as an arc (u, v) from node u to node v in a directed graph, and by this convention the statement 'from $x_j = q$ to $x_j = s$' corresponds to saying 'arc (j, q) replaces arc (j, s)'. A component change of the form 'from $x_j = q$ to $x_j \neq q$' (or 'from $x_j \neq q$ to $x_j = q$') then corresponds to saying that arc (j, q) is dropped from (or added to) the graph. We note it is always possible to encode the pair of conditions $x_j = q$ and $x_j \neq q$ as the assignment of values to a binary variable, e.g. letting $x_{jq} = 1$ denote $x_j = q$ and letting $x_{jq} = 0$ denote $x_j \neq q$, and in the present example this yields the standard algebraic notation for expressing that arc (j, q) is absent or present in a graph.

Broadly speaking, regardless of the representation employed, a move can be determined to be tabu by a restriction defined over any set of conditions on its attributes, provided these attributes are currently tabu-active. As the preceding discussion illustrates, a common type of restriction operates by selecting some subset of attributes and declaring the move to be tabu if a certain minimum number (e.g. one or all) are tabu-active.

3.2.6 Recency-based tabu memory functions

To keep track of the status of move attributes that compose tabu restrictions, and to determine when these restrictions are applicable, several basic kinds of memory functions have been found useful. Two common examples of recency-based memory functions are specified by the arrays *tabu_start*(e) and *tabu_end*(e), where e ranges over attributes relevant to a particular application. These arrays respectively identify the starting and ending iterations of the tabu tenure for attribute e, thus bracketing the period during which e is tabu-active.

The rule for identifying appropriate values for $tabu\_start(e)$ and $tabu\_end(e)$ results from keeping track of the attributes at each iteration that are components of the current move. In particular, on iteration i, if e is an attribute of the current move, and tabu status is defined to avoid reversals, then we set $tabu\_start(\bar{e}) = i + 1$, indicating that the reverse attribute begins its tabu-active status at the start of the next iteration. (For example, if e represents 'from $x_j = p$' then $\bar{e}$ can represent 'to $x_j = p$'.) Attribute $\bar{e}$ will retain this status throughout its tabu tenure, which we denote by t. This then yields $tabu\_end(\bar{e}) = i + t$, so that the tenure for $\bar{e}$ ranges over the t iterations from $i + 1$ to $i + t$.

As a result, it is easy to test whether an arbitrary attribute e is tabu-active, by checking to see if $tabu\_end(e) \geq current\_iteration$. By initializing $tabu\_end(e) = 0$ for all attributes, we insure that $tabu\_end(e) < current\_iteration$, and hence that attribute e is tabu-inactive, until the update previously specified is performed. This suggests we need to keep only the single array $tabu\_end(e)$ to provide information about tabu status. However, we will see that situations arise where it is valuable to keep $tabu\_start(e)$, and either to infer $tabu\_end(e)$ by adding an appropriate value of t (currently computed, or preferably extracted from a pre-stored sequence), or to maintain $tabu\_end(e)$ as a separate array.

Memory can often be further simplified when attributes represent binary alternatives, such as changing from $x_j = 0$ to $x_j = 1$. Then, instead of recording a separate value $tabu\_start(e)$ for each of these attributes, it suffices simply to record a single value $tabu\_start(j)$. We automatically know whether $tabu\_start(j)$ refers to changing from $x_j = 0$ to $x_j = 1$ or the reverse, by taking account of the value of x_j in the current solution. If currently $x_j = 1$, for example, the most recent change was from $x_j = 0$ to $x_j = 1$. Then the reverse attribute, derived from changing x_j from 1 to 0, is the one whose tenure is represented by the value of $tabu\_start(j)$. (We assume that the latest tabu tenure assigned to an attribute takes precedence over all others.)

Regardless of the data structure employed, the key issue for creating tabu status using recency-based memory is to determine a 'good value' of t. Rules for determining t are classified as static or dynamic. *Static rules* choose a value for t that remains fixed throughout the search. *Dynamic rules* allow the value of t to vary. Examples of these two kinds of rules are as follows.

Illustrative Rules to Create Tabu Tenure
(Recency Based)

Static rules	Choose t to be a constant such as $t = 7$ or $t = \sqrt{n}$, where n is a measure of problem dimension.
Dynamic rules	*Simple dynamic*: Choose t to vary (randomly or by systematic pattern) between bounds t_{min} and t_{max}, such as $t_{min} = 5$ and $t_{max} = 11$ or $t_{min} = .9\sqrt{n}$ and $t_{max} = 1.1\sqrt{n}$.
	Attribute-dependent dynamic: Choose t as in the Simple dynamic rule, but determine t_{min} and t_{max} to be larger for attributes that are more attractive, e.g. based on quality or influence considerations.

The indicated values such as 7 and $\sqrt{n}$ are only suggestive, and represent parameters whose preferred values should be set by experimentation for a particular class of problems. Values between 7 and 20 in fact appear to work well for a variety of problem classes, while values between $.5\sqrt{n}$ and $2\sqrt{n}$ appear to work well for other classes. (A weighted multiple of $\sqrt{n}$ is replaced by a weighted multiple of n for some problems.) As previously intimated, if $tabu\_end(e)$ is not maintained separately, but is inferred as the value $tabu\_start(e) + t$, then for the dynamic case it may be preferable to pre-compute a sequence of appropriate values for t and simply step through them each time a new t is needed. (Random sequences can be reasonably approximated this way with considerable saving of computational effort. Alternatively, t can be computed only once or a small number of times on a given iteration, instead of being recomputed separately for each trial move.)

It is often appropriate to allow different types of attributes defining a tabu restriction to be given different values for the tenure t. For example, some attributes can contribute more strongly to a tabu restriction than others, and should be given a briefer tabu tenure to avoid making the restriction too severe. To illustrate, consider a problem of identifying an optimal subset of m items from a much larger set of n items. (For instance, such a problem may involve identifying a subset of m edges from an n-edge graph to create a travelling salesman tour, or a subset of m locations from n available sites to establish distribution centres, or a subset of m nodes from an n-node complex to serve as telecommunication switching centres, etc.) Suppose each move consists of exchanging one or a small number of items in the subset with an equal number outside the subset, to create a new subset of m items. Accompanying this, also suppose a tabu restriction is used that forbids a move if it contains either an item recently added or an item recently dropped, where the tabu tenure provides the meaning of 'recently'.

If the tenure for added and dropped items is the same, the preceding restriction can become very lopsided. In particular, when other factors are equal, preventing items in the subset from being dropped is much more restrictive than preventing items not in the subset from being added, since there are far fewer contained in the subset than contained outside. In addition, preventing elements added to the subset from being dropped for a relatively long time can significantly inhibit available choices; hence the tenure for these elements should be made small by comparison to the tenure for preventing elements dropped from the subset from being added, whether by using static or dynamic rules.

Practical experience indicates that dynamic rules are typically more robust than static rules (see, e.g. Glover *et al.* [5]). Good parameter values for dynamic rules normally range over a wider interval, and produce results comparable or superior to the outcomes produced by static rules. Dynamic rules that depend on both attribute type and quality, where greater tenures are allotted to prevent reversals of attributes that participate in high quality moves, have proved quite effective for difficult problems related to scheduling and routing (Dell'Amico and Trubian [6]; Gendreau *et al.* [7]; Laguna *et al.* [25]). In addition, a class of dynamic rules based on introducing moving gaps in tenure, and another class based on exploiting logical relationships underlying attribute sequences, have recently proved effective (Chakrapani and Skorin-Kapov [9]; Dammeyer and Voss [10]). Dynamic rules for determining tabu tenure are among the aspects of tabu search that deserve more study.

3.2.7 Aspiration criteria

Aspiration criteria are introduced in tabu search to determine when tabu restrictions can be overridden, thus removing a tabu classification otherwise applied to a move. The appropriate use of such criteria can be very important for enabling a TS method to achieve its best performance levels.

Early applications employed only a simple type of aspiration criterion, consisting of removing a tabu classification from a trial move when the move yields a solution better than the best obtained so far. (Such a rule is illustrated in the example of Section 3.2.1.) This criterion remains widely used. However, other aspiration criteria can also prove effective for improving the search.

A basis for one of these criteria arises by introducing the concept of *influence*, which measures the degree of change induced in solution structure or feasibility. (Influence is often associated with the idea of move distance, i.e. a move of greater distance is conceived of as having greater influence—see [5].) This notion can be illustrated for a problem of distributing unequally weighted objects among boxes, where the goal is to give each box as nearly as possible the same weight. A high influence move, which significantly changes the structure of the current solution, is exemplified by a move that transfers a heavy weight object from one box to another, or that swaps objects of very dissimilar weights between two boxes. Such a move may or may not improve the current solution, though it is less likely to yield an improvement when the current solution is relatively good. But high influence moves are important, especially during intervals of breaking away from local optimality, because a series of moves that is confined to making only small structural changes is unlikely to uncover a chance for significant improvement. (Such an effect is illustrated in job sequencing problems by exchanging positions of jobs that are close together.)

Moves of lower influence may normally be tolerated until the opportunities for gain from them appear to be negligible. At such a point, and in the absence of improving moves, aspiration criteria should shift to give influential moves a higher rank. Also, once an influential move is made, tabu restrictions previously established for less influential moves should be dropped or 'weakened', in a manner to be explained. (Bias that may be employed to favour the choice of other influential moves should likewise be temporarily diminished.) These considerations of move influence interact with considerations of regionality and search direction, as indicated below.

Aspirations are of two kinds: *move aspirations* and *attribute aspirations.* A move aspiration, when satisfied, revokes the move's tabu classification. An attribute aspiration, when satisfied, revokes the attribute's tabu-active status. In the latter case the move may or may not change its tabu classification, depending on whether the tabu restriction can be activated by more than one attribute.

The table below lists criteria for determining the admissibility of a trial solution, x^{trial}, as a candidate for consideration (potentially to become x^{next}), where x^{trial} is generated by a move that ordinarily would be classified tabu. The first of these criteria is rarely applicable, but is understood automatically to be part of any tabu search procedure. These aspiration criteria include several useful strategies for tabu search that have not yet been widely examined and that warrant fuller investigation.

For example, a special case of the Regional Aspiration by Objective occurs by defining $R = \{x : g(x) = r\}$, where $g(x)$ is a hashing function created to distinguish among different x vectors according to the value assigned to $g(x)$. (E.g. $g(x)$ can be an integer-valued function defined modulo p, taking the values $r = 0, 1, \ldots, p-1$.) Then *best_cost*(R) is conveniently recorded as *best_cost*(r), identifying the minimum $c(x)$ found when $g(x) = r$. The 'regionality' defined by R in this case provides a basis for integrating the elements of aspiration and differentiation. (A $g(x)$ hashing function can also be treated as an attribute function, and incorporated into tabu restrictions as described earlier. Or in reverse, a hashing function can be defined over attributes, with particular emphasis on those that qualify as influential.) Such an approach can be employed to complement uses of hashing functions in tabu search suggested by Hansen and Jaumard [11] and by Woodruff and Zemel [12].

Aspiration by Search Direction and Aspiration by Influence provide attribute aspirations rather than move aspirations. In most cases attribute and move aspirations are equivalent. (Among the tabu restrictions (R1) to (R7) of Section 3.2.5, only (R3) can provide conditions where these two types of aspirations differ, i.e. where an attribute may be tabu-inactive without necessarily revoking the tabu classification of the associated move.) Nevertheless, different means are employed for testing these two kinds of aspirations.

Illustrative Aspiration Criteria

Aspiration by Default: If all available moves are classified tabu, and are not rendered admissible by some other aspiration criteria, then a 'least tabu' move is selected. (For example, select a move that loses its tabu classification by the least increase in the value of *current_iteration*, or by an approximation to this condition.)

Aspiration by Objective: Global form (customarily used): A move aspiration is satisfied, permitting x^{trial} to be a candidate for selection, if $c(x^{trial}) < best\_cost$.
Regional form: Subdivide the search space into regions $R \in \mathbf{R}$, identified by bounds on values of functions $g(x)$ (or by time intervals of search). Let $best\_cost(R)$ denote the minimum $c(x)$ for x found in R. Then, for $x^{trial} \in R$, a move aspiration is satisfied (for moving to x^{trial}) if $c(x^{trial}) < best\_cost(R)$.

Aspiration by Search Direction: Let $direction(e) = improving$ if the most recent move containing $\overline{e}$ was an improving move, and $direction(e) = nonimproving$, otherwise. ($direction(e)$ and $tabu\_end(e)$ are set to their current values on the same iteration.) An attribute aspiration for e is satisfied (making e tabu-inactive) if $direction(e) = improving$ and the current trial move is an improving move, i.e. if $c(x^{trial}) < c(x^{now})$.

Aspiration by Influence: Let $influence(e) = 0$ or 1 according to whether the move that establishes the value of $tabu\_start(e)$ is a low influence move or a high influence move ($influence(e)$ is set at the same time as setting $tabu\_start(e)$). Also, let $latest(L)$, for $L = 0$ or 1, equal the most recent iteration that a move of influence level L was made. Then an attribute aspiration for e is satisfied if $influence(e) = 0$ and $tabu\_start(e) < latest(1)$. ($e$ is associated with a low influence move, and a high influence move has been performed since establishing the tabu status for e.) For multiple influence levels, $L = 0, 1, 2, \ldots$, the aspiration for e is satisfied if there is an $L > influence(e)$ such that $tabu\_start(e) < latest(L)$.

Aspiration criteria refinements

Refinements of the criteria illustrated above provide an opportunity to enhance the power of tabu search for applications that are more complex, or that offer a large reward for solutions of very high quality. We identify some of the possibilities for achieving this in the following.

Creating a tabu status that varies by degrees, rather than simply designating an attribute to be tabu-active or tabu-inactive, leads to an additional refinement of Aspiration by Search Direction and Aspiration by Influence. Graduated tabu status is implicit in the penalty function and probabilistic variants of tabu search, where status is customarily expressed as a function of how recently or frequently an attribute has become tabu-active. However, to employ this idea to enhance the preceding aspiration criteria, we create a single additional intermediate tabu state that falls between the two states of tabu-active and tabu-inactive. In particular, when an aspiration is satisfied for an attribute that otherwise is tabu-active, we call it a *pending tabu attribute.*

A move that would be classified tabu if its pending tabu attributes are treated as tabu-active, but that would not be classified tabu otherwise, is correspondingly called a *pending tabu move.* A pending tabu move can be treated in one of two ways. In the least restrictive approach, such a move is not prevented from being selected, but is shifted in status so that it will only be a candidate for selection if no improving moves exist except those that are tabu. In the more moderate approach, a pending tabu move additionally must be an improving move to qualify for selection. (This will occur automatically for Aspiration by Search Direction, since in this case a move can only become a pending tabu move when it is improving.)

An Aspiration consequence for Strong Admissibility The preceding notions lead to an additional type of aspiration. Define a move to be strongly admissible if:

(1) it is admissible to be selected and does not rely on aspiration criteria to qualify for admissibility; or

(2) it qualifies for admissibility based on the Global Aspiration by Objective, by satisfying $c(x^{trial}) < best\_cost$.

Aspiration by Strong Admissibility: Let *last_nonimprovement* equal the most recent iteration that a nonimproving move was made, and let *last_strongly_admissible* equal the most recent iteration that a strongly admissible move was made. Then, if *last_nonimprovement* < *last_strongly_admissible*, reclassify every improving tabu move as a pending tabu move (thus allowing it to be a candidate for selection if no other improving moves exist).

The inequality *last_nonimprovement* < *last_strongly_admissible* of the preceding aspiration condition implies two things: first that a strongly admissible improving move has been made since the last nonimproving move, and second that the search is currently generating an improving sequence. (The latter results since only improving moves can occur on iterations more recent than *last_nonimprovement*, and the set of such iterations is nonempty.)

This type of aspiration ensures that the method will always proceed to a local optimum whenever an improving sequence is created that contains at least one strongly admissible move. In fact, condition (2) defining a strongly admissible move can be removed without altering this effect, since once the criterion $c(x^{trial}) < best\_cost$ is used to justify a move selection, then it will continue to be satisfied by all improving moves on subsequent iterations until a local optimum is reached.

Because of its extended ability to override tabu status, the Aspiration by Strong Admissibility may be predicated on the requirement that a move with a high influence level has been made since the end of the most recent (previous) improving sequence. Specifically, such a high influence move should have occurred on an iteration greater than the most recent iteration prior to *last_nonimprovement* on which an improving move was executed. This added requirement is applicable whether or not Aspiration by Influence is used.

These ideas can be used to generate an *alternating TS method* related to the tabu thresholding approach of Glover [13]. Such a method results by adding a further condition to the Aspiration by Strong Admissibility, stipulating that once a nonimproving move is executed, then no improving move is allowed unless it is strongly admissible, thereby generating what may be called an alternating tabu path. The consequence is that each improving sequence in such an alternating tabu path terminates with a local optimum. (An Aspiration by Default must also be considered a strongly admissible move to assure this in exceptional cases.)

The effect of tabu status in this alternating approach can be amplified during a nonimproving phase by interpreting the value *tabu_end*(e) to be shifted to a larger value for all attributes e, until a strongly admissible move is executed and the phase ends. Recent results by Ryan [14] on the depth and width of paths linking local optima are relevant to determining ranges for shifting *tabu_end*(e) in such alternating constructions.

Special considerations for Aspiration by Influence

The Aspiration by Influence criterion can be modified to create a considerable impact on its effectiveness for certain types of applications. The statement of this aspiration derives from the observation that a move characteristically is influential by virtue of containing one or more influential attributes (jobs with large set-up or processing times, warehouses with large capacities, circuits with multiple switches, etc.). Under such conditions, it is appropriate to consider levels of influence defined over attributes, as expressed by *influence(e)*. In other cases, however, a move may derive its influence from the unique combination of attributes involved, and Aspiration by Influence then preferably translates into a move aspiration rather than an attribute aspiration. (In some instances the attribute orientation can be maintained by defining *influence(e)* to be the influence of the trial move that contains e.)

More significantly, in many applications influence depends on a form of connectivity, causing its effects to be expressed primarily over a particular range. We call this range the *sphere of influence* of the associated move or attribute. For example, in the problem of distributing weighted objects among boxes, a move that swaps objects between two boxes has a relatively narrow sphere of influence, affecting only those future moves that transfer an object into, or out of, one of these two boxes. Accordingly, under such circumstances Aspiration by Influence should be confined to modifying the tabu status of attributes, or the tabu classification of moves, that fall within an associated sphere of influence. In the example of swapping objects between boxes, the attributes rendered tabu-inactive would be restricted to *from-attributes* associated with moving an object out of one of the two boxes and *to-attributes* associated with moving an object into one of these boxes. The change of tabu status continues to depend on the conditions noted previously. The influence of the attribute (or move containing it) must be less than that of the earlier move, and the iteration *tabu_start*(e) for the attribute must precede the iteration on which the earlier influential move occurred. These conditions can be registered by setting a flag for *tabu_start*(e) when the influential move is executed, without having to check again later to see if e is affected by such a move. When *tabu_start*(e) becomes reassigned a new value, the flag is dropped.

As the preceding observations suggest, effective measures of move influence and associated characterizations of spheres of influence are extremely important. In addition, it should be noted that influence can be expressed as a function of tabu search memory components, as where a move containing attributes that have neither recently nor frequently been tabu-active may be classified as more highly influential (because executing the move will change the tabu status of these attributes more radically). This encourages a dynamic definition of influence, which varies according to the current search state. These multiple aspects of move influence are likely to constitute a more significant area for future investigation in tabu search.

3.2.8 Frequency-based memory

Frequency-based memory provides a type of information that complements the information provided by recency-based memory, broadening the foundation for selecting preferred moves. Like recency, frequency is often weighted or decomposed into subclasses by taking account of the dimensions of solution quality and move influence.

For our present purposes, we conceive frequency measures to consist of ratios, whose numerators represent counts of the number of occurrences of a particular event (e.g. the number of times a particular attribute belongs to a solution or move) and whose denominators generally represent one of four types of quantities, as shown below.

Denominators for Frequency Measures

(D1)	The total number of occurrences of all events represented by the numerators (such as the total number of associated iterations).
(D2)	The sum of the numerators.
(D3)	The maximum numerator value.
(D4)	The average numerator value.

Denominators (D3) and (D4) give rise to what may be called relative frequencies. The meaning of these different types of frequencies will be clarified by examples below. In cases where the numerators represent weighted counts, some of which may be negative, (D3) and (D4) are expressed as absolute values and (D2) is expressed as a sum of absolute values (possibly shifted by a small constant to avoid a zero denominator).

Let $x(1), x(2), \ldots, x(current\_iteration)$ denote the sequence of solutions generated to the present point of the search process, and let S denote a subsequence of this solution sequence. We take the liberty of treating S as a set as well as an ordered sequence. Elements of S are not necessarily consecutive elements of the full solution sequence. (For example, we sometimes will be interested in cases where S consists of different subsets of high quality local optima.)

Notationally, we let $S(x_j = p)$ denote the set of solutions in S for which $x_j = p$, and let $\#S(x_j = p)$ denote the cardinality of this set (hence the number of times x_j receives the value p over $x \in S$). Similarly, let $S(x_j = p \text{ to } x_j = q)$ denote the set of solutions in S that result by a move that changes $x_j = p$ to $x_j = q$. Finally, let $S(\text{from } x_j = p)$ and $S(\text{to } x_j = q)$ denote the sets of solutions in S that respectively contain $x_j = p$ as a *from-attribute* or $x_j = q$ as a *to-attribute* (for a move to the next solution, or from the preceding solution, in the sequence $x(1), \ldots, x(current\_iteration)$). In general, if *solution_attribute* represents any attribute of a solution that can take the rôle of a *from-attribute* or a *to-attribute* for a move, and if *move_attribute* represents an arbitrary move attribute denoted by (*from-attribute, to-attribute*), then

$$
\begin{aligned}
S(solution\_attribute) &= \{x \in S : x \text{ contains } solution\_attribute\ \} \\
S(move\_attribute) &= \{x \in S : x \text{ results from a move containing } move\_attribute\ \} \\
S(from\text{-}attribute) &= \{x \in S : x \text{ initiates a move containing } from\text{-}attribute\ \} \\
S(to\text{-}attribute) &= \{x \in S : x \text{ results from a move containing } to\text{-}attribute\ \}.
\end{aligned}
$$

The quantity $\#S(x_j = p)$ constitutes a *residence measure*, since it identifies the number of times the attribute $x_j = p$ resides in the solutions of S. Correspondingly, we call a frequency that results by dividing such a measure by one of the denominators (1) to (4) a *residence frequency*. For the numerator $\#S(x_j = p)$, the denominators (1) and (2) both correspond to $\#S$, while denominators (3) and (4) respectively are given by Max $(\#S(x_k = q) : \forall k, q)$ and by Mean $(\#S(x_k = q) : \forall k, q)$.

The quantities $\#S(x_j = p \text{ to } x_j = q)$, $\#S(\text{from } x_j = p)$ and $\#S(\text{to } x_j = q)$ constitute *transition measures*, since they identify the number of times x_j changes from and/or to specified values. Likewise, frequencies based on such measures are called *transition frequencies*. Denominators for creating such frequencies from the foregoing measures include $\#S$, the total number of times the indicated changes occur over S for different j, p and/or q values, and associated Max and Mean quantities.

Distinctions between frequency types

Residence frequencies and transition frequencies sometimes convey related information, but in general carry different implications. They are sometimes confused (or treated identically) in the literature. A noteworthy distinction is that residence measures, by contrast to transition measures, are not concerned with whether a particular solution attribute of an element $x(i)$ in the sequence S is a *from-attribute* or a *to-attribute*, or even whether it is an attribute that changes in moving from $x(i)$ to $x(i+1)$ or from $x(i-1)$ to $x(i)$. It is only relevant that the attribute can be a *from-attribute* or a *to-attribute* in some future move. Such measures can yield different types of implications depending on the choice of the subsequence S.

A high residence frequency, for example, may indicate that an attribute is highly attractive if S is a subsequence of high quality solutions, or may indicate the opposite if S is a subsequence of low quality solutions. On the other hand, a residence frequency that is high (or low) when S contains both high and low quality solutions may point to an entrenched (or excluded) attribute that causes the search space to be restricted, and that needs to be jettisoned (or incorporated) to allow increased diversity.

From a computational standpoint, when S consists of all solutions generated after a specified iteration, then a residence measure can be currently maintained and updated by reference to values of the *tabu_start* array, without the need to increment a set of counters at each iteration. For a set S whose solutions do not come from sequential iterations, however, residence measures are calculated simply by running a tally over elements of S.

Transition measures are generally quite easy to maintain by performing updates during the process of generating solutions (assuming the conditions defining S, and the attributes whose transition measures are sought, are specified in advance). This results from the fact that typically only a few types of attribute changes are considered relevant to track when one solution is replaced by the next, and these can readily be isolated and recorded. The frequencies in the example of Section 3.2.1 constitute an instance of transition frequencies that were maintained in this simple manner. Their use in this example, however, encouraged diversity by approximating the type of rôle that residence frequencies are usually better suited to take.

As a final distinction, a high transition frequency, in contrast to a high residence frequency, may indicate an associated attribute is a 'crack filler', that shifts in and out of solution to perform a fine tuning function. Such an attribute may be interpreted as the opposite of an influential attribute, as considered earlier in the discussion of Aspiration by Influence. In this context, a transition frequency may be interpreted as a measure of volatility.

Examples and uses of frequency measures

Illustrations of both residence and transition frequencies are as follows. (Only numerators are indicated, understanding denominators to be provided by conditions (1) to (4) above.)

Example Frequency Measures (Numerators)

(F1)	$\#S(x_j = p)$
(F2)	$\#S(x_j = p$ for some $x_j)$
(F3)	$\#S(\text{to } x_j = p)$
(F4)	$\#S(x_j$ changes), i.e. $\#S(\text{from-or-to } x_j = p$ for some $p)$
(F5)	$\sum_{x \in S(x_j=p)} c(x)/\#S(x_j = p)$
(F6)	Replace $S(x_j = p)$ in (F5) with $S(x_j \neq p \text{ to } x_j = p)$
(F7)	Replace $c(x)$ in (F6) with a measure of the influence of the solution attribute $x_j = p$

The measures (F5) - (F7) implicitly are weighted measures, created by reference to solution quality in (F5) and (F6), and by reference to move influence in (F7) (or more precisely, influence of an attribute composing a move). Measure (F5) may be interpreted as the average $c(x)$ value over S when $x_j = p$. This quantity can be directly compared to other such averages or can be translated into a frequency measure using denominators such as the sum or maximum of these averages.

Attributes that have greater frequency measures, just as those that have greater recency measures (i.e. that occur in solutions or moves closer to the present), can initiate a tabu-active status if S consists of consecutive solutions that end with the current solution. However, frequency-based memory typically finds its most productive use as part of a longer term strategy, which employs incentives as well as restrictions to determine which moves are selected. In such a strategy, restrictions are translated into evaluation penalties, and incentives become evaluation enhancements, to alter the basis for qualifying moves as attractive or unattractive.

To illustrate, an attribute such as $x_j = p$ with a high residence frequency may be assigned a strong incentive ('profit') to serve as a *from-attribute*, thus resulting in the choice of a move that yields $x_j \neq p$. Such an incentive is particularly relevant in the case where $tabu\_start(x_j \neq p)$ is small, since this value identifies the latest iteration that $x_j \neq p$ served as a *from-attribute* (for avoiding reversals), and hence discloses that $x_j = p$ has been an attribute of every solution since.

Frequency-based memory therefore is usually applied by introducing graduated tabu states, as a foundation for defining penalty and incentive values to modify the evaluation of moves. A natural connection exists between this approach and the recency-based memory approach that creates tabu status as an all-or-none condition. If the tenure of an attribute in recency-based memory is conceived of as a conditional threshold for applying a very large penalty, then the tabu classifications produced by such memory can be interpreted as the result of an evaluation that becomes strongly inferior when the penalties are activated. It is reasonable to anticipate that conditional thresholds should also be relevant to determining the values of penalties and incentives in longer term strategies. Most applications at present, however, use a simple linear multiple of a frequency measure to create a penalty or incentive term. Fundamental ways for taking advantage of frequency based memory are indicated in the next section.

3.2.9 Frequency-based memory in simple intensification and diversification processes

The rôles of intensification and diversification in tabu search are already implicit in several of the preceding prescriptions, but they become especially relevant in longer term search processes. Intensification strategies undertake to create solutions by aggressively encouraging the incorporation of 'good attributes'. In the short term this consists of incorporating attributes receiving highest evaluations by the approaches and criteria previously described, while in the intermediate to long term it consists of incorporating attributes of solutions from selected élite subsets (implicitly focusing the search in subregions defined relative to these subsets). Diversification strategies instead seek to generate solutions that embody compositions of attributes significantly different from those encountered previously during the search. These two types of strategies counterbalance and reinforce each other in several ways.

We first examine simple forms of intensification and diversification approaches that make use of frequency-based memory. These approaches will be illustrated by reference to residence frequency measures, but similar observations apply to the use of transition measures, taking account of contrasting features previously noted.

For a diversification strategy we choose S to be a significant subset of the full solution sequence; for example, the entire sequence starting with the first local optimum, or the subsequence consisting of all local optima. (For certain strategies based on transition measures, S may usefully consist of the subsequence containing each maximum unbroken succession of non-improving moves that immediately follow a local optimum, focusing on $S(to\_attribute)$ for these moves.)

For an intensification strategy we choose S to be a small subset of élite solutions (high quality local optima) that share a large number of common attributes, and secondarily whose members can reach each other by relatively small numbers of moves, independent of whether these solutions lie close to each other in the solution sequence. For example, collections of such subsets S may be generated by clustering procedures, followed by employing a parallel processing approach to treat each selected S separately.

Below we provide rules for generating a penalty or incentive function, PI, which apply equally to intensification and diversification strategies. However, the function PI creates a penalty for one strategy (intensification or diversification) if and only if it creates an incentive for the other. For illustrative purposes, suppose that a move currently under consideration includes two move attributes, denoted e and f, which further may be expressed as $e = (e\_from, e\_to)$ and $f = (f\_from, f\_to)$. To describe the function PI, we let $F(e\_from)$ and $F(e\_to)$ etc. denote the frequency measure for the indicated *from-attributes* and *to-attributes*, and let $T_1, T_2, \ldots, T_6$ denote selected positive thresholds, whose values depend on the case considered.

These conditions for defining PI are related to those previously illustrated to identify conditions in which attributes become tabu-active. For example, specifying that (1) must be positive to make PI positive corresponds to introducing a tabu penalty (or incentive) when both measures exceed their common threshold. If a measure is expressed as the duration since an attribute was most recently made tabu-active, and if the threshold represents a common limit for tabu tenure, then (1) can express a recency-based restriction for determining a tabu classification. Assigning different thresholds to different attributes in (1) corresponds to establishing attribute-dependent tabu tenures. Similarly, the remaining values (2) through (6) may be interpreted as analogues of values that define recency-based measures for establishing a tabu classification, implemented in this case by a penalty.

Illustrative Penalty/Incentive Function PI for To-attributes

Choose PI as a monotonic nondecreasing function of one of the following quantities, where PI is positive when the quantity is positive, and is 0 otherwise. (PI yields a penalty in a diversification strategy and an incentive in an intensification strategy.)

(1) $\text{Min}\{F(e\_to), F(f\_to)\} - T_1$

(2) $\text{Max}\{F(e\_to), F(f\_to)\} - T_2$

(3) $\text{Mean}\{F(e\_to), F(f\_to)\} - T_3$

Illustrative Penalty/Incentive Function PI for From-attributes

Choose PI as a monotonic nondecreasing function of one of the following quantities, where PI is positive when the quantity is positive, and is 0 otherwise. (PI yields an incentive in a diversification strategy and a penalty in an intensification strategy.)

(4) $\text{Min}\{F(e\_from), F(f\_from)\} - T_4$

(5) $\text{Max}\{F(e\_from), F(f\_from)\} - T_5$

(6) $\text{Mean}\{F(e\_from), F(f\_from)\} - T_6$

From these observations, it is clear that the frequency measure F may be extended to represent combined measures of both recency and frequency. Although these measures are already implicitly combined—when penalties and incentives based on frequency measures are joined with tabu classifications based on recency measures, as a foundation for selecting current moves—it is possible that other forms of combination are superior. For example, human problem-solving appears to rely on combinations of these types of memory that incorporate a time-discounted measure of frequency. Such considerations may lead to the design of more intelligent functions for capturing preferred combinations of these memory types.

3.3 Broader Aspects of Intensification and Diversification

Intensification and diversification approaches that use penalties and incentives represent only one class of such strategies. A larger collection emerges by direct consideration of intensification and diversification goals. We examine several approaches that have been demonstrated to be useful in previous applications, and also indicate approaches we judge to have promise in applications of the future. To begin, we make an important distinction between diversification and randomization.

3.3.1 Diversification versus randomization

Seeking a diversified collection of solutions is very different from seeking a randomized collection of solutions. In general, we are interested not just in diversified collections but also in diversified *sequences*, since often the order of examining elements is important in tabu search. This can apply, for example, where we seek to identify a sequence of new solutions (not seen before) so that each successive solution is *maximally diverse* relative to all solutions previously generated. This includes possible reference to a baseline set of solutions, such as $x \in S$, which takes priority in establishing the diversification objective (i.e. where the first level goal is to establish diversification relative to S, and then in turn relative to other solutions generated). The diversification concept applies as well to generating a diverse sequence of numbers or a diverse set of points from the vertices of a unit hypercube.

Let $Z(k) = \{z(1), z(2), \ldots, z(k)\}$ represent a sequence of points drawn from a set Z. For example, Z may be a line interval if the points are scalars. We take $z(1)$ to be a seed point of the sequence. (The seed point may be excepted from the requirement of belonging to Z.) Then we define $Z(k)$ to be a *diversified sequence* (or simply a *diverse sequence*), relative to a chosen distance metric d over Z by requiring each subsequence $Z(h)$ of $Z(k), h < k$, and each associated point $z = z(h+1)$ to satisfy the following hierarchy of conditions:

(A) z maximizes the minimum distance $d(z, z(i))$ for $i \leq h$;

(B) subject to (A), z maximizes the minimum distance $d(z, z(i))$ for $1 < i \leq h$, then for $2 < i \leq h, \ldots$, etc. (in strict priority order);

(C) subject to (A) and (B), z maximizes the distance $d(z, z(i))$ for $i = h$, then for $i = h - 1, \ldots$, and finally for $i = 1$. (Additional ties may be broken arbitrarily.)

To handle diversification relative to an initial baseline set Z^* (such as a set of solutions $x \in S$), the preceding hierarchy of conditions is preceded by a condition stipulating that z first maximizes the minimum distance $d(z, z^*)$ for $z^* \in Z^*$. A useful (weaker) variant of this condition simply treats points of Z^* as if they constitute the last elements of the sequence $Z(h)$.

Variations on (A), (B), and (C), including going deeper in the hierarchy before arbitrary tie breaking, are evidently possible. Such conditions make it clear that a diverse sequence is considerably different from a random sequence. Further, they are computationally very demanding to satisfy. Even by omitting condition (B), and retaining only (A) and (C), if the elements $z(i)$ refer to points on a unit hypercube, then by our present state of knowledge the only way to generate a diverse sequence of more than a few points is to perform comparative enumeration. (However, a diverse sequence of points on a line interval, particularly if $z(1)$ is an endpoint or midpoint of the interval, can be generated with much less difficulty.) Because of this, it can sometimes be useful to generate sequences by approximating the foregoing conditions (see Glover [15]). Taking a broader view, an extensive effort to generate diverse sequences can be performed in advance, independent of problem solving efforts, so that such sequences are pre-computed and available as needed. Further, a diverse sequence for elements of a high dimensional unit hypercube may be derived by reverse projection techniques ('lifting' operations) from a sequence for a lower dimensional hypercube, ultimately making reference to sequences from a line interval.

Biased diversification, just as biased random sampling, is possible by judicious choices of the set Z. Also, while the goals of diversification and randomization are somewhat different, the computational considerations share a feature in common. To generate a random sequence by the strict definition of randomness would require massive effort. Years of study have produced schemes for generating sequences that empirically approximate this goal, and perhaps a similar outcome may be possible for generating diversified sequences. The hypothesis of tabu search, in any case, is that recourse to diversification is more appropriate (and more powerful) in the problem solving context than recourse to randomization.

We note these observations can be applied in a setting, as subsequently discussed, where the device of producing a solution 'distant from' another is accomplished not by reference to a standard distance metric, but rather by a series of displacements which involve selecting a move from a current neighbourhood at each step. (In this case the metric may derive from differences in weighted measures defined over *from-attributes* and *to-attributes*.) An application of these ideas is given in Kelly *et al.* [16], and we also discuss a special variation under the heading of 'Path Relinking' below. This stepwise displacement approach is highly relevant to those situations where neighbourhood structures are essential for preserving desired properties (such as feasibility).

3.3.2 Reinforcement by restriction

One of the early types of intensification strategy, characterized in terms of exploiting strongly determined and consistent variables in Glover [17], begins by selecting a set S as indicated for determining a penalty and incentive function, i.e. one consisting of élite solutions grouped by a clustering measure. Instead of (or in addition to) creating penalties and incentives, with the goal of incorporating attributes into the current solution that have high frequency measures over S, the method of reinforcement by restriction operates by narrowing the range of possibilities allowed for adding and dropping such attributes. For example, if $x_j = p$ has a high frequency over S for only a small number of values of p, then moves are restricted to allow x_j to take only one of these values in defining a *to-attribute*. Thus, if x_j is a 0-1 variable with a high frequency measure over S for one of its values, this value will become fixed once an admissible move exists that allows such a value assignment to be made. Other assignments may be permitted, by a variant of Aspiration by Default, if the current set of restricted alternatives is unacceptable.

Initial consideration suggests such a restriction approach offers nothing beyond the options available by penalties and incentives. However, the approach can accomplish more than this for two reasons. First, explicit restrictions can substantially accelerate the execution of choice steps by reducing the number of alternatives examined. Second, and more significantly, many problems simplify and collapse once a number of explicit restrictions are introduced, allowing structural implications to surface that permit these problems to be solved far more readily.

Reinforcement by restriction is not limited to creating an intensification effect. Given finite time and energy to explore alternatives, imposing restrictions on some attributes allows more variations to be examined for remaining unrestricted attributes than otherwise would be possible. Thus, intensification with respect to selected elements can enhance diversification over other elements, creating a form of selective diversification. Such diversification may be contrasted with the exhaustive diversification created by the more rigid memory structures of branch and bound. In an environment where the finiteness of available search effort is dwarfed by the number of alternatives that exist to be explored exhaustively, selective diversification can make a significant contribution to effective search.

Path relinking

Path relinking is initiated by selecting two solutions x' and x'' from a collection of élite solutions produced during previous search phases. A path is then generated from x' to x'', producing a solution sequence $x' = x'(1), x'(2), \ldots, x'(r) = x''$, where $x'(i+1)$ is created from $x'(i)$ at each step by choosing a move that leaves the fewest number of moves remaining to reach x''. (A choice criterion for approximating this effect is indicated below.) Finally, once the path is completed, one or more of the solutions $x'(i)$ is selected as a solution to initiate a new search phase.

This approach provides a fundamental means for pursuing the goals of intensification and diversification when its steps are implemented to exploit strategic choice rule variations. A number of alternative moves will typically qualify to produce a next solution from $x'(i)$ by the 'fewest remaining moves' criterion, consequently allowing a variety of possible paths from x' to x''. Selecting unattractive moves relative to $c(x)$ at each step will tend to produce a final series of strongly improving moves, while selecting attractive moves will tend to produce lower quality moves at the end. (The last move, however, will be improving, or leave $c(x)$ unchanged, since x'' is a local optimum.) Thus, choosing best, worst or average moves, using an aspiration criterion to override choices in the last two cases if a sufficiently attractive solution is available, provide options that produce contrasting effects in generating the indicated sequence. (Arguments exist in favour of selecting best moves at each step, and then repeating the process by interchanging x' and x''.)

The issue of an appropriate aspiration is more broadly relevant to selecting a preferred $x'(i)$ for launching a new search phase, and to terminating the sequence early. The choice of one or more solutions $x'(i)$ to launch a new search phase preferably should depend not only on $c(x'(i))$ but also on the values $c(x)$ of those solutions x that can be reached by a move from $x'(i)$. In particular, when $x'(i)$ is examined to move to $x'(i+1)$, a number of candidates for $x = x'(i+1)$ will be presented for consideration. The process additionally may be varied to allow solutions to be evaluated other than those that yield $x'(i+1)$ closer to x''.

Let $x^*(i)$ denote a neighbour of $x'(i)$ that yields a minimum $c(x)$ value during an evaluation step, excluding $x^*(i) = x'(i+1)$. (If the choice rules do not automatically eliminate the possibility $x^*(i) = x'(h)$ for $h < i$, then a simple tabu restriction can be used to do this.) Then the method selects a solution $x^*(i)$ that yields a minimum value for $c(x^*(i))$ as a new point to launch the search. If only a limited set of neighbours of $x'(i)$ are examined to identify $x^*(i)$, then a superior least cost $x'(i)$, excluding x' and x'', may be selected instead. Early termination may be elected upon encountering an $x^*(i)$ that yields $c(x^*(i)) < \mathrm{Min}\{c(x'), c(x''), c(x'(p))\}$, where $x'(p)$ is the minimum cost $x'(h)$ for all $h \leq i$. (The procedure continues without stopping if $x'(i)$, in contrast to $x^*(i)$, yields a smaller $c(x)$ value than x' and x'', since $x'(i)$ effectively adopts the rôle of x'.)

Variation and tunnelling

A variant of the path relinking approach proposed in Glover [15] starts both endpoints x' and x'' simultaneously, producing two sequences $x' = x'(1), \ldots, x'(r)$ and $x'' = x''(1), \ldots, x''(s)$. The choices are designed to yield $x'(r) = x''(s)$, for final values of r and s. To progress toward this outcome when $x'(r) \neq x''(s)$, either $x'(r)$ is selected to create $x'(r+1)$, by the criterion of minimizing the number of moves remaining to reach $x''(s)$, or $x'(s)$ is chosen to create $x''(s+1)$, by the criterion of minimizing the number of moves remaining to reach $x'(r)$. From these options, the move is selected that produces the smallest $c(x)$ value, thus also determining which of r or s is incremented on the next step.

The path relinking approach can benefit by a tunnelling approach that allows a different neighbourhood structure to be used than in the standard search phase. In particular, it often is desirable periodically to allow moves for path relinking that would normally be excluded due to the creation of infeasibility. Such a practice is less susceptible to becoming 'lost' in an infeasible region than other ways of allowing periodic infeasibility, since feasibility evidently must be recovered by the time x'' is reached. The tunnelling effect thus created offers a chance to reach solutions that might otherwise be bypassed. In the variant that starts from both x' and x'', at least one of $x'(r)$ and $x''(s)$ may be kept feasible.

Path relinking can be organized to place greater emphasis on intensification or diversification by choosing x' and x'' to share more or fewer attributes in common. Similarly choosing x' and x'' from a clustered set of élite solutions will stimulate intensification, while choosing them from two widely separated sets will stimulate diversification.

3.3.3 Extrapolated relinking

An extension of the path relinking approach, which we call extrapolated relinking, goes beyond the path endpoint x'' (or alternatively x'), to obtain solutions that span a larger region. The ability to continue beyond this endpoint results from a method for approximating the move selection criterion specified for the standard path relinking approach, which seeks a next solution that leaves the fewest moves remaining to reach x''.

Specifically, let $A(x)$ denote the set of solution attributes in x, and let A^{drop} denote the set of solution attributes that are dropped by moves performed to reach the current solution $x'(i)$, i.e. the attributes that have served as *from-attributes* in these moves. (Some of these may have been reintroduced into $x'(i)$, but they also remain in A^{drop}.) Then we seek a move at each step to maximize the number of *to-attributes* that belong to $A(x'') - A(x'(i))$, and subject to this to minimize the number that belong to $A^{drop} - A(x'')$. Such a rule can generally be implemented very efficiently, by data structures limiting the examination of moves to those containing *to-attributes* of $A(x'') - A(x'(i))$ (or permitting these moves to be examined before others).

Once $x'(r) = x''$ is reached, the process continues by modifying the choice rule as follows. The criterion now selects a move to maximize the number of its *to-attributes* not in A^{drop} minus the number of its *to-attributes* that are in A^{drop}, and subject to this, to minimize the number of its *from-attributes* that belong to $A(x'')$. (The combination of these criteria establishes an effect analogous to that achieved by the standard algebraic formula for extending a line segment beyond an endpoint. However, the secondary minimization criterion is probably less important.) The path then stops whenever no choice remains that permits the maximization criterion to be positive.

For neighbourhoods that allow relatively unrestricted choices of moves, this approach yields an extension beyond x'' that introduces new attributes, without re-incorporating any old attributes, until no move remains that satisfies this condition. The ability to go beyond the limiting points x' and x'' creates a form of diversification not available to the path that 'lies between' these points. At the same time the exterior points are influenced by the trajectory that links x' and x''.

3.3.4 Solutions evaluated but not visited

Intensification and diversification strategies may profit by the fact that a search process generates information not only about solutions actually visited, but also about additional solutions evaluated during the examination of moves not taken. One manifestation of this is exploited by reference to the solutions $x^*(i)$ in the path relinking approach.

From a different point of view, let S^* denote a subset of solutions evaluated but not visited (for example, taken from the sequence $x(1), \ldots, x(current\_iteration)$) whose elements x yield $c(x)$ values within a chosen band of attractiveness. It is relatively easy to maintain a count such as $\#S^*(\text{to } x_j = p)$, which identifies the number of times $x_j = p$ is a *to-attribute* of a trial move leading to a solution of S^*. Such a count may be differentiated further, by stipulating that the trial move must be improving, and of high quality relative to other moves examined on the same iteration. (Differentiation of this type implicitly shrinks the composition of S^*.) Then an attribute that achieves a relatively high frequency over S^*, but that has a low residence frequency over solutions actually visited, is given an incentive to be incorporated into future moves, simultaneously serving the goals of both intensification and diversification. Recency and frequency interact in this approach, by disregarding the incentive if the attribute has been selected on a recent move.

3.3.5 Interval-specific penalties and incentives

A useful adjunct to the preceding ideas extends the philosophy of Aspiration by Search Direction and Aspiration by Strong Admissibility. By these aspiration criteria, improving moves are allowed to escape a tabu classification under certain conditions, but with the result of lowering their status so that they are treated as inferior improving moves.

An extension of this preserves the improving/non-improving distinction when penalties and incentives are introduced that are not intended to be pre-emptive. For this extension, evaluations are again divided into the intervals of improving and non-improving. Penalties and incentives are then given limited scope, degrading or enhancing evaluations within a given interval, but without altering the relationship between evaluations that lie in different intervals.

Incentives granted on the basis of influence are similarly made subject to this restricted shift of evaluation. Since an influential move is not usually improving in the vicinity of a local optimum, maintaining the relationship between evaluations in different intervals implies such moves will usually be selected only when no improving moves exist, other than those classified tabu. But influential moves also have a recency-based effect. Just as executing a high influence move can cancel the tabu classification of a lower influence move over a limited span of iterations, so it should reduce or cancel the incentive to select other influential moves for a corresponding duration.

3.3.6 Candidate list procedures

Section 3.2.4 stressed the importance of procedures to isolate a candidate subset of moves from a large neighbourhood, to avoid the computational expense of evaluating moves from the entire neighbourhood. Procedures of this form have been used in optimization methods for almost as long as issues of reducing computational effort have been taken seriously (since at least the 1950s and probably much earlier). Some of the more strategic forms of these procedures came from the field of network optimization (Glover *et al.* [18], Mulvey [19], Frendewey [20]). In such approaches, the candidate subset of moves is referenced by a list that identifies their defining elements (such as indexes of variables, nodes, or arcs), and hence these approaches have acquired the name of *candidate list strategies.*

A simple form of candidate list strategy is to construct a single element list by sampling from the neighbourhood space at random, and to repeat the process if the outcome is deemed unacceptable. This is the foundation of Monte Carlo methods, as noted earlier. Studies from network optimization, however, suggest that approaches based on more systematic designs produce superior results. Generally, these involve decomposing a neighbourhood into critical subsets, and using a rule that assures subsets not examined on one iteration become scheduled for examination on subsequent iterations. For subsets appropriately determined, best outcomes result by selecting highest quality moves from these subsets, either by explicit examination of all alternatives or by using an adaptive threshold to identify such moves (see Glover *et al.* [21]).

Another kind of candidate list strategy periodically examines larger portions of the neighbourhood, creating a master list of some number of best alternatives found. The master list is then consulted to identify moves (derived from or related to those recorded) for additional iterations until a threshold of acceptability triggers the creation of a new master list.

Candidate list strategies implicitly have a diversifying influence by causing different parts of the neighbourhood space to be examined on different iterations. This suggests there may be benefit from co-ordinating such strategies with other diversification strategies, an area that remains open for investigation.

Candidate list strategies also lend themselves very naturally to parallel processing, where forms of neighbourhood decomposition otherwise examined serially are examined in parallel. Moves can be selected by choosing the best candidate from several processes, or instead each process can execute its own preferred move, generating parallel solution trajectories that are periodically co-ordinated at a higher level. These latter approaches hold considerable promise. Some of the options are described in Glover *et al.* [5].

3.3.7 Compound neighbourhoods

Identifying an effective neighbourhood for defining moves from one solution to another can be extremely important. For example, an attempt to solve a linear programming problem by choosing moves that increment or decrement problem variables, versus choosing moves that use pivot processes or directional search, obviously can make a substantial difference to the quality of the final solution obtained. The innovations that have made linear programming a powerful optimization tool rely significantly on the discovery of effective neighbourhoods for making moves.

For combinatorial applications where the possibilities for creating neighbourhoods are largely confined to various constructive or destructive processes, or to exchanges, improvements often result by combining neighbourhoods to create moves. For example, in sequencing applications such as that illustrated in Section 3.2.1, it is generally preferable to combine neighbourhoods consisting of insert moves and swap moves, allowing both types of moves to be considered at each step. Another way of combining neighbourhoods is to generate compound moves, where a sequence of simpler moves is treated as a single more complex move.

A special type of approach for creating compound moves results from a succession of steps in which an element is assigned to a new state, with the outcome of *ejecting* some other element from its current state. The ejected element is then assigned to a new state, in turn ejecting another element, and so forth, creating a chain of such operations. For example, such a process occurs in a job sequencing problem by moving a job to a new position occupied by another job, thereby ejecting this job from its position. The second job is then moved to a new position to eject another job, and so on, finally ending by inserting the last ejected job between two other jobs. This type of approach, called an *ejection chain strategy*, has useful applications for problems of many types, particularly in connection with scheduling, routing, and partitioning (Glover [22, 23], Dorndorf and Pesch [24]). A tabu search method incorporating this approach has proved highly successful for multilevel generalized assignment problems (Laguna *et al.* [25]), suggesting the relevance of ejection chain strategies for creating compound neighbourhoods in other tabu search applications.

3.3.8 Creating new attributes—vocabulary building and concept formation

A frontier area of tabu search involves the creation of new attributes out of others. The learning approach called *target analysis*, which can implicitly combine or subdivide attributes to yield a basis for improved move evaluations has been effectively used in conjunction with tabu search in scheduling applications (see Section 3.4), and provides one of the means for generating new attributes. We focus here, however, on creating new attributes by reference to a process that may be called *vocabulary building*, related to concept formation.

Vocabulary building is based on viewing a chosen set S of solutions as a text to be analyzed, by undertaking to discover attribute combinations shared in common by various solutions x in $\mathbf{X}$. Attribute combinations that emerge as significant enough to qualify as units of vocabulary, by a process to be described below, are treated as new attributes capable of being incorporated into tabu restrictions and aspiration conditions. In addition, they can be directly assembled into larger units as a basis for constructing new solutions.

We represent collections of attributes by encoding them as assignments of values to variables, which we denote by $y_j = p$, to differentiate the vector y from the vector x which possibly may have a different dimension and encoding. Normally we suppose a y vector contains enough information to be transformed into a unique x, to which it corresponds, but this assumption can be relaxed to allow more than one x to yield the same y. (It is to be noted that a specified range of different assignments for a given attribute can be expressed as a single assignment for another, which is relevant to creating vocabulary of additional utility.)

Let $Y(S)$ denote the collection of y vectors corresponding to the chosen set S of x vectors. In addition to assignments of the form $y_j = p$ which define attributes, we allow each y_j to receive the value $y_j = *$, in order to generate subvectors that identify specific attribute combinations. In particular, an attribute combination will be implicitly determined by the non-$*$ values of y.

The approach to generate vocabulary units will be to compare vectors y' and y'' by an intersection operator, $Int(y', y'')$ to yield a vector $z = Int(y', y'')$ by the rule: $z_j = y'_j$ if $y'_j = y''_j$, and $z_j = *$ if $y'_j \neq y''_j$. By this definition we also obtain $z_j = *$ if either y'_j or

$y''_j = *$. Int is associative, and the intersection $Int(y : y \in \mathbf{Y})$, for an arbitrary $\mathbf{Y}$, yields a z in which $z_j = y_j$ if all y_j have the same value for $y \in \mathbf{Y}$, and $z_j = *$ otherwise.

Accompanying the intersection operator, we also define a relation of containment, by the stipulation that y'' *contains* y' if $y'_j = *$ for all j such that $y'_j \neq y''_j$. Associated with this relation, we identify the *enclosure* of y' (relative to S) to be the set $Y(S : y') = \{y \in Y(S) : y$ contains $y'\}$, and define the enclosure value of y', $enc\_value(y')$, to be the number of elements in this set, i.e. the value $\#Y(S : y')$. Finally, we refer to the number of non-$*$ components of y' as the *size* of the vector, denoted $size(y')$. (If $y \in Y(S)$, the size of y is the same as its dimension.)

Clearly the greater $size(y')$ becomes, the smaller $enc\_value(y')$ tends to become. Thus for a given size s, we seek to identify vectors y' with $size(y') \geq s$ that maximize $enc\_value(y')$, and for a given enclosure value v to identify vectors y' with $enc\_value(y') \geq v$ that maximize $size(y')$. Such vectors are included among those regarded as qualifying as vocabulary units.

Similarly we include reference to weighted enclosure values, where each $y \in Y(S)$ is weighted by a measure of attractiveness (such as the value $c(x)$ of an associated solution $x \in S$), to yield $enc\_value(y')$ as a sum of the weights over $Y(S : y')$. Particular attribute values may likewise be weighted, as by a measure of influence, to yield a weighted value for $size(y')$, equal to the sum of weights over non-$*$ components of y'.

From a broader perspective, we seek vectors as vocabulary units that give rise to aggregate units called *phrases* and *sentences* with certain properties of consistency and meaning, characterized as follows. Each y_j is allowed to receive one additional value, $y_j = blank$, which may be interpreted as an empty space free to be filled by another value (in contrast to $y_j = *$, which may be interpreted as a space occupied by two conflicting values). We begin with the collection of vectors created by the intersection operator Int, and replace the $*$ values with $blank$ values in these vectors. We then define an extended intersection operator E_Int, where $z = E\_Int(y', y'')$ is given by the rules defining Int if y'_j and y''_j are not blank. Otherwise $z_j = y'_j$ if $y''_j = blank$, and $z_j = y''_j$ if $y'_j = blank$. E_Int is likewise associative. The vector $z = E\_Int(y : y \in Y)$ yields $z_j = *$ if any two $y \in Y$ have different non-*blank* values y_j, or if some y has $y_j = *$. Otherwise z_j is the common y_j value for all y with y_j non-*blank* (where $z_j = blank$ if $y_j = blank$ for all y).

The y vectors created by *E_Int* are those we call *phrases.* A *sentence* (implicitly, a complete sentence) is a phrase that has no *blank* values. We call a phrase or sentence *grammatical* (logically consistent) if it has no $*$ values. Thus grammatical sentences are y vectors lacking both blank values and $*$ values, constructed from attribute combinations (subvectors) derived from the original elements of $Y(S)$. Finally we call a grammatical sentence y *meaningful* if it corresponds to, or maps into, a feasible solution x. (Sentences that are not grammatical do not have a form that permits them to be translated into an x vector, and hence cannot be meaningful.)

The elements of $Y(S)$ are all meaningful sentences, assuming they are obtained from feasible x vectors, and the goal is to find other meaningful sentences obtained from grammatical phrases and sentences constructed as indicated. More precisely, we are interested in generating meaningful sentences (hence feasible solutions) that are not limited to those that can be obtained from $Y(S)$, but that can also be obtained by one of the following strategies:

Sentence Construction Strategies

(S1) Translate a grammatical phrase into a sentence by filling in the blanks (by the use of neighbourhoods that incorporate constructive moves).

(S2) Identify some set of existing meaningful sentences (e.g. derived from current feasible x vectors not in S), and identify one or more phrases, generated by *E_Int* over S, that lie in each of these sentences. Then, by a succession of moves from neighbourhoods that preserve feasibility, transform each of these sentences into new meaningful sentences that retain as much of the identified phrases as possible.

(S3) Identify portions of existing meaningful sentences that are contained in grammatical phrases, and transform these sentences into new meaningful sentences (using feasibility preserving neighbourhoods) by seeking to incorporate additional components of the indicated phrases.

The foregoing strategies can be implemented by incorporating the same tabu search incentive and penalty mechanisms for choosing moves indicated in previous sections. We assume in these strategies that neighbourhood operations on x vectors are directly translated into associated changes in y vectors. In the case of (S1) there is no assurance that a meaningful sentence can be achieved unless the initial phrase itself is meaningful (i.e. is contained in at least one meaningful sentence) and the constructive process is capable of generating an appropriate completion. Also, in (S3) more than one grammatical phrase can contain a given part (subvector) of a meaningful sentence, and it may be appropriate to allow the targeted phrase to change according to possibilities consistent with available moves.

Although we have described vocabulary building processes in a somewhat general form to make their range of application visible, specific instances can profit from special algorithms for linking vocabulary units into sentences that are both meaningful and attractive, in the sense of creating good $c(x)$ values. An example of this is provided by vocabulary building approaches for the travelling salesman problem described in [23], where vocabulary units can be transformed into tours by specialized shortest path procedures. A number of combinatorial optimization problems are implicit in generating good sentences by these approaches, and the derivation of effective methods for handling these problems in various settings, as in the case of the travelling salesman problem, may provide a valuable contribution to search procedures generally.

3.3.9 Strategic oscillation

The strategic oscillation approach is closely linked to the origins of tabu search, and provides an effective interplay between intensification and diversification over the intermediate to long term. Strategic oscillation operates by moving until hitting a boundary, represented by feasibility or a stage of construction, that normally would represent a point where the method would stop. Instead of stopping, however, the neighbourhood definition is extended, or the evaluation criteria for selecting moves is modified, to permit the boundary to be crossed. The approach then proceeds for a specified depth beyond the boundary, and turns around. At this point the boundary is again approached and crossed, this time from the opposite direction, proceeding to a new turning point. The process of repeatedly approaching and crossing the boundary from different directions creates a form of oscillation that gives the method its name. Control over this oscillation is established by generating modified evaluations and rules of movement, depending on the region currently navigated and the direction of search. The possibility of retracing a prior trajectory is avoided by standard tabu mechanisms.

A simple example of this approach occurs for the multidimensional knapsack problem, where values of 0-1 variables are changed from 0 to 1 until reaching the boundary of feasibility. It then continues into the infeasible region using the same type of changes, but with a modified evaluator. After a selected number of steps, direction is reversed by changing variables from 1 to 0. Evaluation criteria to drive toward improvement (or smallest disimprovement) vary according to whether the movement is from more-to-less or less-to-more feasible (or infeasible), and are accompanied by associated restrictions on admissible changes to values of variables. An implementation of such an approach by Freville and Plateau [26, 27] has generated particularly high quality solutions for multidimensional knapsack problems.

A somewhat different type of application occurs for the problem of finding an optimal spanning tree subject to inequality constraints on subsets of weighted edges. One type of strategic oscillation approach for this problem results from a constructive process of adding edges to a growing tree until it is spanning, and then continuing to add edges to cross the boundary defined by the tree construction. A different graph structure results when the current solution no longer constitutes a tree, and hence a different neighbourhood is required, yielding modified rules for selecting moves. The rules again change in order to proceed in the opposite direction, removing edges until again recovering a tree. In such problems, the effort required by different rules may make it preferable to cross a boundary to different depths on different sides. One option is to approach and retreat from the boundary while remaining on a single side, without crossing (i.e. electing a crossing of 'zero depth'). In this example, additional types of boundaries may be considered, derived from the inequality constraints.

The use of strategic oscillation in applications that alternate constructive and destructive processes can be accompanied by exchange moves that maintain the construction at a given level. A ***proximate optimality principle***, which states roughly that good constructions at one level are likely to be close to good constructions at another, motivates a strategy of applying exchanges at different levels, on either side of a target structure such as a spanning tree, to obtain refined constructions before proceeding to adjacent levels.

Finally, we remark that the boundary incorporated in strategic oscillation need not be defined in terms of feasibility or structure, but can be defined in terms of a region where the search appears to gravitate. The oscillation then consists of compelling the search to move out of this region and allowing it to return.

3.4 Tabu Search Applications

Tabu search is still in an early stage of development, with a substantial majority of its applications occurring only since 1989. However, TS methods have enjoyed successes in a variety of problem settings, as represented by the partial list shown in the table below. Scheduling provides one of the most fruitful areas for modern heuristic techniques in general and for tabu search in particular. Although the scheduling applications presented in Table 3.1 are limited to those found in the published literature (or about to appear), there are a number of studies currently in progress that deal with scheduling models corresponding to modern manufacturing systems.

One of the early applications of TS in scheduling is due to Widmer and Hertz [28], who develop a TS method for the solution of the permutation flow shop problem. This problem consists of n multiple operation jobs arriving at time zero to be processed in the same order on m continuously available machines. The processing time of a job on a given machine is fixed (deterministic) and individual operations are not pre-emptable. The objective is to find the ordering of jobs that minimizes the *makespan*—the completion time of the last job.

Widmer and Hertz use a simple insertion heuristic based on a travelling salesman analogy to the permutation flow shop problem to generate the starting ordering of the jobs. The procedure considers neighbourhoods defined by swap moves, and at each iteration the best non-tabu move is executed evaluated relative to $c(x)$. The tabu tenure is exclusively set to a value of 7 moves and the tabu restriction is based on the paired attributes (job index, position). The termination criterion is specified as a maximum number of iterations.

Table 3.1: Some applications of tabu search

Brief Description	Reference
Scheduling	
Employee scheduling	Glover & McMillan [61]
Flow shop	Widmer & Hertz [28] Taillard [30] Reeves [31]
Job shop with tooling constraints	Widmer [32]
Convoy scheduling	Bovet *et al.* [62]
Single machine scheduling	Laguna *et al.* [8]
Just-in-time scheduling	Laguna & Gonzalez-Velarde [63]
Multiple-machine weighted flow time	Barnes & Laguna [64]
Flexible-resource job shop	Daniels & Mazzola [33]
Job shop scheduling	Dell'Amico & Trubian [6]
Single machine (target analysis)	Laguna & Glover [3]
Resource scheduling	Mooney & Rardin [36]
Deadlines and setup times	Woodruff & Spearman [37]
Transportation	
Travelling salesman	Malek *et al.* [38] Glover [22]
Vehicle routing	Gendreau *et al.* [39] Osman [42] Semet & Taillard [43]
Layout and circuit design	
Quadratic assignment	Skorin-Kapov [65] Taillard [41] Chakrapani & Skorin-Kapov [9]
Electronic circuit design	Bland & Dawson [66]
Telecommunications	
Path assignment	Oliveira & Stroud [67] Anderson *et al.* [58]
Bandwidth packing	Glover & Laguna [59]
Graphs	
Clustering	Glover *et al.* [68] Hansen *et al.* [45]
Graph colouring	Hertz & de Werra [69] Hertz *et al.* [70]
Stable sets in large graphs	Friden *et al.* [71]
Maximum clique	Gendreau *et al.* [7]
Probabilistic logic and expert systems	
Maximum satisfiability	Hansen & Jaumard [11]
Probabilistic logic	Jaumard *et al.* [44]
Probabilistic logic/expert systems	Hansen *et al.* [60]
Neural networks	
Learning in an associative memory	de Werra & Hertz [56]
Nonconvex optimization problems	Beyer & Ogier [57]
Others	
Multiconstraint 0-1 knapsack	Dammeyer & Voss [10]
Large-scale controlled rounding	Kelly *et al.* [47]
General fixed charge	Sun & McKeown [72]

Computational experiments compare this TS implementation with six previously developed heuristic methods. The study examines 50 problems with n and m ranging from values of 5 to 20 where the maximum number of TS iterations is set to $n + m$. In direct competition with the best previous heuristic developed by Nawaz *et al.* [29], the TS method returns superior solutions for 58%, and matches the best solution found for 92% of the problems.

This early TS procedure does not include many of the mechanisms described in this chapter which are now established as important components of the more effective procedures. Nevertheless, the study was important for being one of the first of its type, and for disclosing the relevance of TS for scheduling, thus motivating other research to follow in this area.

The study of Taillard [30] is noteworthy in this regard, applying tabu search to the flow shop sequencing problem. This work demonstrates that tabu search obtains solutions uniformly better than the best of the classical heuristics, while investing comparable solution time. In addition, although optimality of the solutions could not be proved, by allowing sufficient CPU time Taillard's TS method found optimal solutions for every problem for which a such solution was known. Reeves [31] further improves the computational efficiency of this method by incorporating a candidate list strategy; using this approach, TS consistently outperformed a simulated annealing heuristic on a wide variety of problem instances. Another study in this area by Widmer [32] develops a TS method for the solution of an important problem in scheduling models for flexible manufacturing—the job shop scheduling problem with tooling constraints. This implementation establishes the ability of the TS approach to be adapted to handle highly complex problems, with practical features disregarded by previous studies of related problems reported in the literature.

Daniels and Mazzola [33] present a TS method for the flexible-resource flow shop scheduling problem, which generalizes the classic flow shop scheduling problem by allowing job-operation processing times to depend on the amount of resource assigned to an operation. The objective is to determine the job sequence, resource-allocation policy, and operation start times that optimize system performance. The TS method employs a nested-search strategy based on a decomposition of the problem into these three main components (job sequencing, resource allocation, and operation start times). The procedure was tested on over 1600 problems and is reported to be extremely effective. On 480 problem instances small enough to permit optimal solutions to be identified, the TS approach obtained optimal solutions for over 70% of the test problems, while incurring an average error of 0.3% and a maximum deviation from optimality of 2.5%. On larger problems, comparisons with other heuristic procedures showed the TS method was able to find significantly superior solutions. In addition, the authors note the nested TS approach holds considerable promise for efficient implementation in a parallel processing setting.

Dell'Amico and Trubian [6] apply tabu search to the notoriously difficult job-shop scheduling problem. They develop a *bi-directional* method to find 'good' feasible starting solutions. Their procedure alternates between assigning operations at the beginning and at the end of a partial schedule, which contrasts with previous uni-directional List Scheduler algorithms. In addition to starting from a good solution, their TS procedure assigns tabu tenures that are dependent on the search state and are selected from a given range. The range is periodically revised using uniform distributions to determine new upper and lower bounds. A simple intensification strategy is used that recovers the best solution found so far and treats it as the current solution, when a given number of iterations have been performed without improving the best solution. Computational experiments with 53 benchmark problem instances show this TS method is highly robust, in contrast to previously published local search procedures for this problem. In particular, the TS method outperforms two simulated annealing methods due to van Laarhoven *et al.* [34] and Matsuo *et al.* [35] in terms of both solution quality and speed. In addition, Dell'Amico and Trubian establish new best solutions for five out of seven open problems in the literature.

Laguna and Glover [3] develop a tailored TS method for the solution of a class of single machine scheduling problems with delay penalties and setup costs. This research discloses the usefulness of target analysis as a means of integrating effective diversification strategies within tabu search. The study also establishes the importance of accounting for regional dependencies of good decision criteria. The resulting procedure obtains solutions that are uniformly as good as, or better than, the best previously known solutions over a wide variety of problem instances. For large problems (with 100 jobs) the margin of superiority of the method is more dramatic. (The previously best available heuristic for this class of problems was also a TS procedure, as empirically shown by Laguna *et al.* [8].)

Mooney and Rardin [36] develop a TS procedure for a special case of the problem of assigning tasks to a single primary resource, subject to constraints resulting from the pre-assignment of secondary or auxiliary resources. Potential applications of this problem include shift-oriented production and manpower scheduling problems and course scheduling, where classrooms may be primary and instructors and students may be secondary resources. This study includes 7 variants of a basic TS procedure. These variants combine the use of deterministic and random candidate list construction, several move selection rules, and strategic oscillation. An index is created to measure the level of diversification that each variant of the method is capable of achieving. Extensive experiments with randomly-generated and real data show that the TS variants with strategic oscillation achieve high levels of diversification (as measured by the defined index) while outperforming alternative approaches. The motivation for measuring diversification levels stems from the authors' conjecture that 'an algorithm that diversifies the search must cover the search space more or less evenly'. As a result of this study, it was found that a simple iterated descent approach (see Section 3.2.3) obtained high diversification levels but performed poorly in terms of solution quality. Therefore, relatively high diversification appears to be a necessary but not a sufficient condition for finding good solutions.

Woodruff and Spearman [37] present a highly innovative TS procedure for production scheduling, addressing a general sequencing problem that includes two classes of jobs with setup times, setup costs, holding costs and deadlines. A TS method is used with insertion moves to transform one trial solution into another. Due to the presence of deadline constraints, not every sequence is feasible. However, the search path is allowed to visit infeasible solutions by a form of strategic oscillation. A candidate list is also used as a means of reducing the computational effort involved in evaluating a given neighbourhood. Diversification is achieved by introducing a parameter d into the cost function. Low values of d result in the selection of the best available move (with reference to the objective function value) as customarily done in a deterministic tabu search, while high values result in a randomized move selection which resembles a variant of probabilistic tabu search.

The tabu list designed for this approach is based on the concept of hashing functions. The list is composed of two entries for each visited sequence, the cost and the value of a simple hashing function (i.e. a value that represents the ordering of jobs in the sequence). Computational experiments were conducted on simulated data that captured the characteristics of the demand and production environment in a large circuit board plant. For a set of twenty test problems, the average deviation from optimality was 3% and optimal solutions were achieved in seventeen cases. The best solutions were found during searches using d values other than zero, which supports the contention that long-term memory considerations become important in complex problem settings. This study also marks the first application of TS where hashing functions are used to control the tabu structures. A more detailed study on these kinds of functions and their use within the TS framework is given in Woodruff and Zemel [12].

The first parallel implementation of tabu search to appear in the literature is due to Malek *et al.* [38]. In this implementation, each child process runs a copy of a serial TS method with different parameter settings (i.e. tabu list size and tabu restrictions). After specified intervals, the child processes are halted and the main process compares their results. The main process then selects the 'best' solution found and gives it as the initial solution to all the child processes. The 'best' solution is generally the one with the least tour cost, but an alternative solution is passed if the tour has been used before. The tabu data structures are blanked every time that the child processes are temporarily stopped. This scheme requires little overhead due to interprocessor communication, and implements an intensification phase around 'good' solutions that is not easily reproduced in a serial environment. This research shows the importance of parallel computing in solving large combinatorial optimization problems, and it also illustrates one possibility for exploiting the flexibility of tabu search in this kind of environment. Joining such an approach with the use of stronger move neighbourhoods, such as those of Gendreau *et al.* [39] or of Glover [40], may be expected to yield additional improvements.

Chakrapani and Skorin-Kapov [9] present a parallel implementation on the Connection Machine CM-2 of a TS method for the quadratic assignment problem. The implementation uses n^2 processors, where n is the size of the problem. A moving gap strategy is used to vary the tabu tenure dynamically. Additional intensification and diversification are achieved via frequency-based memory. The procedure proves to be very effective in terms of solution quality. The largest problems that can currently be solved by exact methods are of size $n = 20$. The authors' method easily matches all known optimal solutions and also matches best known solutions for additional problems of size up to $n = 80$. (These solutions were obtained by a TS procedure due to Taillard [41].) In addition the study by Chakrapani and Skorin-Kapov reports new best solutions to a set of published problems of size $n = 100$. A careful implementation on the Connection Machine, a massively parallel system, proves to be extremely suitable in this context. The increase in time per iteration appears to be a logarithmic function of n. This study also offers directions for alternative implementations that may be more efficient when solving very large quadratic assignment problems.

Vehicle routing constitutes another important area with many practical applications. Several TS variants and a hybrid simulated annealing/TS approach for the vehicle routing problem under capacity and distance constraints are presented by Osman [42]. The neighbourhoods are defined using a so-called λ-interchange. The hybrid simulated annealing approach, which uses a nonmonotonic TS strategy for adjusting temperatures, improves significantly over a standard SA. The hybrid approach produces new best solutions for 7 instances in a set of 14 previously published problems. However, this approach exhibits a large variance with regard to solution quality and computational time. The pure TS methods also find 7 new best solutions to problems in the same set, and in addition they maintain a good average solution quality without excessive computational effort. The procedures developed by Osman are easily adapted to the vehicle routing problem with different vehicle sizes.

Gendreau *et al.* [39] also develop a TS procedure for vehicle routing, using a somewhat different move neighbourhood than used in [42]. Their approach is tested against the previously reigning best solution approaches in the literature, and outperforms all of them in most problems. Interestingly, in spite of the different choice of move neighbourhoods, their results are quite closely comparable to those of Osman [42].

Semet and Taillard [43] address a difficult version of the vehicle routing problem with many complicating side conditions, including different vehicle types and sizes, different regions, and restricted delivery windows. Their outcomes improve significantly over those previously obtained for those problems, and again demonstrate the ability of tabu search to be adapted to handle diverse real world features.

One of the first TS methods to use more than one tabu list is due to Gendreau *et al.* [7], which is designed to solve the maximum clique problem in graphs. The method uses add-delete moves to define neighbourhoods for the current solutions and a tabu list to store the indexes of the vertices most recently deleted. A second list is used to record the solutions visited during a specified number of most recent iterations. The second list is always active while the first one is only consulted when 'augmenting' moves are considered (i.e. moves that increase the size of the current clique). Storing previously visited solutions as part of the tabu structure is unusual in TS methods, but was achieved in this instance due to cleverly designed data structures to exploit the neighbourhood definition. Multiple tabu lists have now become common in many TS applications.

Jaumard *et al.* [44] investigate the problem of determining the consistency of probabilities that specify whether given collections of clauses are true, with extensions to include probability intervals, conditional probabilities, and perturbations to achieve satisfiability. By integrating a tabu search approach with an exact 0-1 nonlinear programming procedure for generating columns of a master linear program, they readily solved problems with up to 140 variables and 300 clauses, approximately tripling both the number of variables and the number of clauses that could be handled by existing alternative approaches.

This work is extended in the study of Hansen *et al.* [70] to address problems arising in expert systems, as in systems for medical diagnosis. Tabu search is again embedded in a column generation scheme to determine optimal changes to sets of rules that incorporate probabilities. The combinatorial complexity of this problem comes from the fact that the number of columns grows astronomically as a function of the number of logical sentences used to define rules. This extended study is able to generate optimal solutions for rule systems containing up to 200 sentences, significantly advancing the size of such problems that previously could be addressed.

Dammeyer and Voss [10] studied the multiconstrained 0-1 knapsack problem using a TS method that incorporates tabu restrictions based on the logical structure of the attribute sequence generated. The method is compared against an improved version of a simulated annealing method from the literature specifically designed for these problems, using a testbed of 57 problems with known optimal solutions. The TS and SA methods take comparable time on these problems, but the TS method finds optimal solutions for nearly 50% more problems than simulated annealing (44 problems versus 31). On the remaining problems, deviations from optimality with the TS method were less than 2% in all cases, and less than 1% for all problems except one. Dammeyer and Voss also note the SA method to be very sensitive to the choice of control parameters, which greatly influences the solution quality. By contrast, they found the TS parameters to be very robust. Similar differences in outcomes are established in the study of quadratic semi-assignment problems by Domschke *et al.* [46].

When publishing tabular data, the United States Bureau of the Census must sometimes round fractional data to integer values or round integer data to multiples of a pre-specified base. Data integrity can be maintained by rounding tabular data subject to additivity constraints while minimizing the overall perturbation of the data. Kelly *et al.* [47] describe a tabu search procedure with strategic oscillation for solving this NP-hard problem. A lower bound is obtained by solving a network flow programming model and the corresponding solution is used as the starting point for the procedure. Strategic oscillation plays a major rôle in this TS implementation. The oscillation in this case is around the feasibility boundary. A penalty function is used first to lead the search from the lower bound solution towards the feasible region, by linearly incrementing the penalty for an aggregated measure of constraint violation. Once the procedure reaches feasibility for the first time, the penalty oscillates within a specified period. The theoretical lower bound value obtained by network optimization (which may not be attainable by any feasible solution) is used to gauge the quality of solutions found. Experiments with 270 simulated problems yield an average deviation from this lower bound of 1.32%. In addition, for 248 three-dimensional tables provided by the United States Bureau of the Census, the deviation from the lower bound was only 0.391%.

3.5 Connections and conclusions

Relationships between tabu search and other procedures like simulated annealing and genetic algorithms provide a basis for understanding similarities and contrasts in their philosophies, and for creating potentially useful hybrid combinations of these approaches. We offer some speculation on preferable directions in this regard, and also suggest how elements of tabu search can add a useful dimension to neural network approaches.

3.5.1 Simulated annealing

The contrasts between simulated annealing and tabu search are fairly conspicuous, though undoubtedly the most prominent is the focus on exploiting memory in tabu search that is absent from simulated annealing. The introduction of this focus entails associated differences in search mechanisms, and in the elements on which they operate.

Accompanying the differences directly attributable to the focus on memory, and also magnifying them, several additional elements are fundamental for understanding the relationship between the methods. We consider three such elements in order of increasing importance.

First, tabu search emphasizes scouting successive neighbourhoods to identify moves of high quality, as by candidate list approaches of the form described in Section 3.3. This contrasts with the simulated annealing approach of randomly sampling among these moves to apply an acceptance criterion that disregards the quality of other moves available. (Such an acceptance criterion provides the sole basis for sorting the moves selected in the SA method.) The relevance of this difference in orientation is accentuated for tabu search, since its neighbourhoods include linkages based on history, and therefore yield access to information for selecting moves that is not available in neighbourhoods of the type used in simulated annealing.

Next, tabu search evaluates the relative attractiveness of moves not only in relation to objective function change, but also in relation to factors of influence. Both types of measure are significantly affected by the differentiation among move attributes, as embodied in tabu restrictions and aspiration criteria, and in turn by relationships manifested in recency, frequency, and sequential interdependence (hence, again, involving recourse to memory). Other aspects of the state of search also affect these measures, as reflected in the altered evaluations of strategic oscillation, which depend on the direction of the current trajectory and the region visited.

Finally TS emphasizes guiding the search by reference to multiple thresholds, reflected in the tenures for tabu-active attributes and in the conditional stipulations of aspiration criteria. This may be contrasted to the simulated annealing reliance on guiding the search by reference to the single threshold implicit in the temperature parameter. The treatment of thresholds by the two methods compounds this difference between them. Tabu search varies its threshold nonmonotonically, reflecting the conception that multidirectional parameter changes are essential to adapt to different conditions, and to provide a basis for locating alternatives that might otherwise be missed. This contrasts with the simulated annealing philosophy of adhering to a temperature parameter that only changes monotonically.

Hybrids are now emerging that are taking preliminary steps to bridge some of these differences, particularly in the realm of transcending the simulated annealing reliance on a monotonic temperature parameter. A hybrid method that allows temperature to be strategically manipulated, rather than progressively diminished, has been shown to yield improved performance over standard SA approaches, as noted in the work by Osman [42]. Another hybrid method that expands the SA basis for move evaluations has also been found to perform better than standard simulated annealing in the study by Kassou [48].

Consideration of these findings invites the question of whether removing the memory scaffolding of tabu search and retaining its other features may yield a viable method in its own right. A foundation for doing this by a 'tabu thresholding method' is described by Glover [13], and is reported in a study of graph layout and design problems by Verdejo and Cunquero [49] to perform more effectively than previously best methods for these problems.

3.5.2 Genetic algorithms

Genetic algorithms offer a somewhat different set of comparisons and contrasts with tabu search. As will be described in chapter 4, GAs are based on selecting subsets (usually pairs) of solutions from a population, called parents, and combining them to produce new solutions called children. Rules of combination to yield children are based on the genetic notion of crossover, which consists of interchanging solution values of particular variables, together with occasional operations such as random value changes. Children that pass a survivability test, probabilistically biased to favor those of superior quality, are then available to be chosen as parents of the next generation. The choice of parents to be matched in each generation is based on random or biased random sampling from the population (in some parallel versions executed over separate subpopulations whose best members are periodically exchanged or shared). Genetic terminology customarily refers to solutions as chromosomes, variables as genes, and values of variables as alleles.

By means of coding conventions, the genes of genetic algorithms may be compared to attributes in tabu search, or more precisely to attributes in the form underlying the residence measures of frequency-based memory. Introducing memory in GAs to track the history of genes and their alleles over subpopulations would provide an immediate and natural way to create a hybrid with TS.

Some important differences between genes and attributes should be noted, however. Differentiation of attributes into *from* and *to* components, each having different memory functions, do not have a counterpart in genetic algorithms. This results because GAs are organized to operate without reference to moves (although, strictly speaking, combination by crossover can be viewed as a special type of move). Another distinction derives from differences in the use of coding conventions. Although an attribute change, from a state to its complement, can be encoded in a zero-one variable, such a variable does not necessarily provide a convenient or useful representation for the transformations provided by moves. Tabu restrictions and aspiration criteria handle the binary aspects of complementarity without requiring explicit reference to a zero-one x vector or two-valued functions. Adopting a similar orientation (relative to the special class of moves embodied in crossover) might yield benefits for genetic algorithms in dealing with issues of genetic representation, which currently pose difficult questions (see e.g. Liepens and Vose [50]).

A domain where a genetic interpretation of tabu search ideas seems possible concerns the use of vocabulary building approaches, as described in Section 3.3. Vocabulary units may suggestively be given the alternative name of 'genetic material'. By this means, such units may be viewed as substrings of genes, created by a process that selectively extracts them to establish a substring pool. As elements are accumulated from different sources within such a pool, and progressively re-integrated into *phrases* and *sentences* by vocabulary processes, a genetic parallel may be conceived of as incorporating substring templates to guide construction of new genes.

Perhaps the use of such evolving substring pools, as opposed to the exclusive focus on parents and children, would prove useful in genetic algorithms. But there are limiting factors, since the TS processes for creating vocabulary are based on conscious and strategic reconstruction, and hence do not much resemble genetic processes. To preserve the genetic metaphor, one may imagine relying on *intelligent enzymes*, operating as special subroutines to cut out appropriate components and then recombine them according to systematic principles. If this is not stretching analogy too far, the outcome may qualify as an interesting hybrid of the GA and TS approaches.

A contrast to be noted between genetic algorithms and tabu search arises in the treatment of context, i.e. in the consideration given to structure inherent in different problem classes. For tabu search, context is fundamental, embodied in the interplay of attribute definitions and the determination of move neighbourhoods, and in the choice of conditions to define tabu restrictions. Context is also implicit in the identification of amended evaluations created in association with longer-term memory, and in the regionally-dependent neighbourhoods and evaluations of strategic oscillation.

At the opposite end of the spectrum, GA literature characteristically stresses the freedom of its rules from the influence of context. Crossover, in particular, is a *context-neutral* operation, which assumes no reliance on conditions that solutions must obey in a particular problem setting, just as genes make no reference to the environment as they follow their encoded instructions for recombination (except, perhaps, in the case of mutation). Practical application, however, generally renders this an inconvenient assumption, making solutions of interest difficult to find. Consequently, a good deal of effort in GA implementation is devoted to developing 'special crossover' operations that compensate for the difficulties created by context, effectively re-introducing it on a case by case basis. The related branch of evolutionary algorithms does not rely on the narrower genetic orientation, and hence does not regard the provision for context as a deviation (or extra-genetic innovation). Still, within these related families of approaches, there is no rigorous dedication to exploiting context, as manifested in problem structure, and no prescription to indicate how solutions might be combined systematically to achieve such exploitation, with the exception of special problems such as the TSP (see, for instance, the discussion of the paper by Whitley *et al.* [51] in chapter 4).

The chief method by which modern genetic algorithms and their cousins handle structure is by relegating its treatment to some other method. That is, genetic algorithms combine solutions by their parent-children processes at one level, and then a descent method takes over to operate on the resulting solutions to produce new solutions. These new solutions in turn are submitted to be recombined by the GA processes. In these versions, pioneered by Mühlenbein *et al.* [52], and also advanced by Davis [53] and Ulder *et al.* [54], genetic algorithms already take the form of hybrid methods. Hence, as will be further remarked in chapter 4, there is a natural basis for marrying GA and TS procedures in such approaches. But genetic algorithms and tabu search can also be joined in a more fundamental way.

Specifically, tabu search strategies for intensification and diversification are based on the following question: how can information be extracted from a set of good solutions to help uncover additional (and better) solutions? From one point of view, GAs provide an approach for answering this question, consisting of putting solutions together and interchanging components (in some loosely defined sense, if traditional crossover is not strictly enforced). Tabu search, by contrast, seeks an answer by using processes that specifically incorporate neighbourhood structures into their design.

Augmented by historical information, neighbourhood structures are used as a basis for applying penalties and incentives to induce attributes of good solutions to become incorporated into current solutions. Consequently, although it may be meaningless to interchange or otherwise incorporate a set of attributes from one solution into another in a wholesale fashion, as attempted in recombination operations, a stepwise approach to this goal through the use of neighbourhood structures is entirely practicable. This observation, formulated from a slightly different perspective in Glover [15], provides a basis for creating structured combinations of solutions that embody desired characteristics such as feasibility. The use of these structured combinations makes it possible to integrate selected subsets of solutions in any system that satisfies three basic properties. Instead of being compelled to create new types of crossover to remove deficiencies of standard operators upon being confronted by changing contexts, this approach addresses context directly and makes it an essential part of the design for generating combinations. (A related manifestation of this theme is provided by the path relinking approach of Section 3.3.) The current trend of genetic algorithms seems to be increasingly compatible to adopting such an approach, particularly in the work of Mühlenbein [55], and this could provide a basis for a significant hybrid combination of genetic algorithm and tabu search ideas. In particular, we note that Mühlenbein has likewise indicated the relevance of incorporating TS types of memory into GAs.

3.5.3 Neural networks

Neural networks have a somewhat different set of goals from tabu search, although some overlaps exist. We indicate how tabu search can be used to extend certain neural net conceptions, yielding a hybrid that may have both hardware and software implications.

The basic transferable insight from tabu search is that memory components with dimensions such as recency and frequency can increase the efficacy of a system designed to evolve toward a desired state. We suggest there may be merit in fusing neural network memory with tabu search memory. (A rudimentary acquaintance with neural network ideas is assumed.)

Recency-based considerations can be introduced from tabu search into neural networks by a time delay feedback loop from a given neuron back to itself (or from a given synapse back to itself, by the device of interposing additional neurons). This permits firing rules and synapse weights to be changed only after a certain time threshold, determined by the length of the feedback loop. Aspiration thresholds of the form conceived in tabu search can be embodied in inputs transmitted on a secondary level, giving the ability to override the time delay for altering firing thresholds and synaptic weights. Frequency-based effects employed in tabu search may similarly be incorporated by introducing a form of cumulative averaged feedback.

Time delay feedback mechanisms for creating recency and frequency effects can also have other functions. In a problem-solving context, for example, it may be convenient to disregard one set of options to concentrate on another, while retaining the ability to recover the suppressed options after an interval. This familiar type of human activity is not a customary part of neural network design, but can be introduced by the time dependent functions previously indicated. In addition, a threshold can be created to allow a suppressed option to 'go unnoticed' if current activity levels fall in a certain range, effectively altering the interval before the option re-emerges for consideration. Neural network designs to incorporate those features may directly make use of the TS ideas that have made these elements effective in the problem-solving domain.

Tabu search strategies that introduce longer term intensification and diversification concerns are also relevant to neural network processes. As a foundation for blending these approaches, it is useful to adopt an orientation where a collection of neurons linked by synapses with various activation weights is treated as a set of attribute variables which can be assigned alternative values. Then the condition that synapse j (from a specified origin neuron to a specified destination neuron) is assigned an activation weight in interval p can be coded by the assignment $y_j = p$, where y_j is a component of an attribute vector y, as identified in the discussion of attribute creation processes in Section 3.2.5. A similar coding identifies the condition under which a neuron fires (or does not fire) to activate its associated synapses. As a neural network process evolves, a sequence of these attribute vectors is produced over time. The association between successive vectors may be imagined to operate by reference to a neighbourhood structure implicit in the neural architecture and associated connection weights. There may also be an implicit association with some (unknown) optimization problem, or a more explicit association with a known problem and set of constraints. In the latter case, attribute assignments (neuron firings and synapse activation) can be evaluated for efficacy by transformation into a vector x, to be checked for feasibility by $x \in \mathbf{X}$. (We maintain a distinction between y and x since there may not be a one-one association between them.)

Time records identifying the quality of outcomes produced by recent firings, and identifying the frequency with which particular attribute assignments produce the highest quality firing outcomes, yield a basis for delaying changes in certain weight assignments and for encouraging changes in others. The concept of influence, in the form introduced in tabu search, should be considered in parallel with quality of outcomes.

Attribute creation and vocabulary building strategies as discussed in Section 3.3 have a significant potential for contributing to the issue of adaptive network design. An element notably lacking in neural networks at present is a systematic means to generate *concepts*, as where a chess player evolves an ability to detect and treat a particular configuration (class of positions) as a single unit. Vocabulary building yields a direct way to generate new units from existing ones. Applied to neural networks, such a process may operate to find embedded configurations of states that correspond to good firing outcomes, and assemble them into larger units. More particularly, starting with a set of previous firing states and weightings, represented by assignments in which y ranges over a set $Y(S)$, attribute creation processes can be used to identify and integrate significant components (subvectors). Copying and segregating these components permits associated neural connections to be treated as hardwired, i.e. locked in. This corresponds to treating the unit as a single new attribute. Activating the unit (as by setting $y_j = p$ for appropriate j and p) thus automatically activates the full associated system of firings. The duplication of components of y segregated from the original structure permits the 'original components' to continue to evolve without the hardwiring limitation. This occurs in the same way that created attributes in vocabulary building processes exist side by side with separate instances of the attributes that gave rise to them.

As noted in Table 3.1 of Section 3.4, elements of tabu search have already been incorporated into neural networks in the work of de Werra and Hertz [56] and Beyer and Ogier [57]. These applications, which respectively treat visual pattern identification and nonconvex optimization, are reported significantly to reduce training times and increase the reliability of outcomes generated. In addition, TS principles also have been integrated into a special variant of neural networks making use of constructions called ghost images in [40].

The preceding observations suggest that TS concepts and strategies offer a variety of fruitful possibilities for creating hybrid methods in combination with other approaches. Beyond this, many opportunities exist to expand the frontiers of tabu search itself. We have undertaken to point out some of the areas likely to yield particular benefits. As shown in Section 3.4, TS appears to be opening the door to new advances in many settings, encompassing production scheduling, routing, design, network planning, expert systems, and a variety of other areas. Tabu search methods present opportunities for future research both in developing new applications and in creating improved methodology. The exploration of these realms may afford a chance to make a useful impact on the solution of practical combinatorial problems.

Acknowledgement

This work was supported in part by the Joint Air Force Office of Scientific Research and Office of Naval Research Contract No. F49620-90-C-0033 at the University of Colorado.

References

[1] F.Glover (1986) Future paths for integer programming and links to artificial intelligence. *Computers & Ops.Res.*, **5**, 533-549.

[2] P.Hansen (1986) The steepest ascent mildest descent heuristic for combinatorial programming. *Congress on Numerical Methods in Combinatorial Optimization*, Capri, Italy.

[3] M.Laguna and F.Glover (1992) Integrating target analysis and tabu search for improved scheduling systems. *Expert Systems with Applications: An International Journal*, (to appear).

[4] U.Faigle and W.Kern (1992) Some convergence results for probabilistic tabu search. *ORSA J. on Computing*, **4**, 32-37.

[5] F.Glover, E.Taillard and D.de Werra (1993) A user's guide to tabu search. *Annals of Ops.Res.*, **41**, (to appear).

[6] M.Dell'Amico and M.Trubian (1993) Applying tabu search to the job-shop scheduling problem. *Annals of Ops.Res.*, **41**, (to appear).

[7] M.Gendreau, L.Salvail and P.Soriano (1992) Solving the maximum clique problem using a tabu search approach. *Discrete Appl.Math.*, (to appear).

[8] M.Laguna, J.W.Barnes and F.Glover (1991) Tabu search methods for a single machine scheduling problem. *J. of Intelligent Manufacturing*, **2**, 63-74.

[9] J.Chakrapani and J.Skorin-Kapov (1993) Massively parallel tabu search for the quadratic assignment problem. *Annals of Ops.Res.*, **41**, (to appear).

[10] F.Dammeyer and S.Voss (1993) Dynamic tabu list management using the reverse elimination method. *Annals of Ops.Res.*, **41**, (to appear).

[11] P.Hansen and B.Jaumard (1990) Algorithms for the maximum satisfiability problem. *Computing*, **44**, 279-303.

[12] D.L.Woodruff and E.Zemel (1993) Hashing vectors for tabu search *Annals of Ops.Res.*, **41**, (to appear).

[13] F.Glover (1992) *Simple tabu thresholding in optimization.* Graduate School of Business and Administration, University of Colorado at Boulder.

[14] J.Ryan (1992) *Depth and width of local optima.* Department of Mathematics, University of Colorado at Denver.

[15] F.Glover (1992) Tabu search for nonlinear and parametric optimization (with links to genetic algorithms) *Discrete Appl.Math.*, (to appear).

[16] J.P.Kelly, M.Laguna and F.Glover (1992) A study of diversification strategies for the quadratic assignment problem. *Computers & Ops.Res.*, (to appear).

[17] F.Glover (1977) Heuristics for integer programming using surrogate constraints. *Dec.Sci.*, **8**, 156-166.

[18] F.Glover, D.Karney, D.Klingman and A.Napier (1974) A computational study on start procedures, basis change criteria, and solution algorithms for transportation problems. *Man.Sci.*, **20**, 793-813.

[19] J.Mulvey (1978) Pivot strategies for primal simplex network codes. *J. of the ACM*, **25**, 266-270.

[20] J.Frendewey (1983) *Candidate list strategies for GN and Simplex SON methods.* Graduate School of Business and Administration, University of Colorado at Boulder.

[21] F.Glover, R.Glover and D.Klingman (1986) The threshold assignment algorithm. *Math.Prog. Study*, **26**, 12-37.

[22] F.Glover (1992) Multilevel tabu search and embedded search neighbourhoods for the travelling salesman problem. *ORSA J. on Computing.* (to appear).

[23] F.Glover (1992) *Ejection chains, reference structures, and alternating path methods for the travelling salesman problem.* Graduate School of Business and Administration, University of Colorado at Boulder.

[24] U.Dorndorf and E.Pesch (1992) *Fast clustering algorithms.* INFORM and University of Limberg.

[25] M.Laguna, J.P.Kelly, J.L.Gonzalez-Velarde and F.Glover (1991) *Tabu search for the multilevel generalized assignment problem.* Graduate School of Business and Administration, University of Colorado at Boulder.

[26] A.Freville and G.Plateau (1986) Heuristics and reduction methods for multiple constraint 0-1 linear programming problems. *EJOR*, **24**, 206-215.

[27] A.Freville and G.Plateau (1990) Hard 0-1 multiknapsack test problems for size reduction methods. *Investigacion Operativa*, **1**, 251-270.

[28] M.Widmer and A.Hertz (1989) A new heuristic method for the flow shop sequencing problem. *EJOR*, **41**, 186-193.

[29] M.Nawaz, E.E.Emscore, Jr. and I.Ham (1983) A heuristic algorithm for the m-machine, n-job flow-shop sequencing problem. *OMEGA*, **11**, 91-95.

[30] E.Taillard (1990) Some efficient heuristic methods for the flow shop sequencing problem. *EJOR*, **47**, 65-74.

[31] C.R.Reeves (1993) Improving the efficiency of tabu search for machine sequencing problems. *JORS*, (to appear).

[32] M.Widmer (1991) Job shop scheduling with tooling constraints: a tabu search approach. *JORS*, **42**, 75-82.

[33] R.L.Daniels and J.B.Mazzola (1993) A tabu search heuristic for the flexible-resource flow shop scheduling problem. *Annals of Ops.Res.*, **41**, (to appear).

[34] P.J.M.van Laarhoven, E.H.L.Aarts and J.K.Lenstra (1988) *Job shop scheduling by simulated annealing.* OS-R8809, Centre for Mathematics and Computer Science, Amsterdam.

[35] H.Matsuo, C.J.Suh and R.S.Sullivan (1988) *A controlled search simulated annealing method for the general job shop scheduling problem.* Graduate School of Business, The University of Texas at Austin.

[36] E.L.Mooney and R.L.Rardin (1993) Tabu search for a class of scheduling problems. *Annals of Ops.Res.*, **41**, (to appear).

[37] D.L.Woodruff and M.L.Spearman (1992) Sequencing and batching for two classes of jobs with deadlines and setup times. *J. of the Production and Operations Management Society*, (to appear).

[38] M.Malek, M.Guruswamy, M.Pandya and H.Owens (1989) Serial and parallel simulated annealing and tabu search algorithms for the travelling salesman problem. *Annals of Ops.Res.*, **21**, 59-84.

[39] M.Gendreau, A.Hertz and G.Laporte (1992) A tabu search heuristic for the vehicle routing problem. *Man.Sci.*, (to appear).

[40] F.Glover (1992) Optimization by ghost image processes in neural networks. *Computers & Ops.Res.*, (to appear).

[41] E.Taillard (1991) Robust taboo search for the quadratic assignment problem. *Parallel Computing*, **17**, 443-455.

[42] I.H.Osman (1993) Metastrategy simulated annealing and tabu search algorithms for the vehicle routing problem. *Annals of Ops.Res.*, **41**, (to appear).

[43] F.Semet and E.Taillard (1993) Solving real-life vehicle routing problems efficiently using taboo search. *Annals of Ops.Res.*, **41**, (to appear).

[44] B.Jaumard, P.Hansen and M.Poggi di Aragao (1991) Column generation methods for probabilistic logic. *ORSA J. on Computing*, **3**, 135-148.

[45] P.Hansen, B.Jaumard and Da Silva (1992) Average linkage divisive hierarchical clustering. *J. of Classification*, (to appear).

[46] W.Domschke, P.Frost and S.Voss (1991) Tabu search techniques for the quadratic semi-assignment problem. *In* G.Fandel, T.Gulledge and A.Jones (Eds.) *New Directions for Operations Research in Manufacturing*, 389-405. Springer.

[47] J.P.Kelly, B.L.Golden and A.A.Assad (1993) Large-scale rounding using tabu search with strategic oscillation. *Annals of Ops.Res.*, **41**, (to appear).

[48] I.Kassou (1992) *Amelioration d'ordonnancements par des methodes de voisinage.* Doctoral thesis, INSA, Rouen, France.

[49] V.V.Verdejo and R.M.Cunquero (1992) *An application of the tabu thresholding techniques: minimization of the number of arcs crossing in an acyclic digraph.* Departamento de Estadistica e Investigacion Operativa, Universidad de Valencia, Spain.

[50] G.Liepins and M.D.Vose (1990) Representational issues in genetic optimization. *J. of Experimental and Theoretical Artificial Intelligence*, **2**, 101-115.

[51] D.Whitley, T.Starkweather and D.Shaner (1991) The traveling salesman and sequence scheduling: quality solutions using genetic edge recombination. *In* [53], 350-372.

[52] H.Mühlenbein,M.Gorges-Schleuter and O.Krämer (1988) Evolution algorithms in combinatorial optimization. *Parallel Computing*, **7**, 65-85.

[53] L.Davis (Ed.) (1991) *Handbook of Genetic Algorithms.* Van Nostrand Reinhold, New York.

[54] N.Ulder, E.Pesch, P.J.M.van Laarhoven, H.J.Bandelt and E.H.L.Aarts (1991) Genetic local search algorithm for the travelling salesman problem. *In* R.Maenner and H.P.Schwefel (Eds.) *Parallel Problem-solving from Nature.* Lecture Notes in Computer Science 496, Springer-Verlag, 109-116.

[55] H.Mühlenbein (1992) Parallel genetic algorithms in combinatorial optimization. *In* O.Balci (Ed.) *Computer Science and Operations Research.*, Pergamon Press.)

[56] D.de Werra and A.Hertz (1989) Tabu search techniques: a tutorial and an application to neural networks. *OR Spektrum*, **11**, 131-141.

[57] D.Beyer and R.Ogier (1991) Tabu learning: a neural network search method for solving nonconvex optimization problems. *Proceedings of the International Joint Conference on Neural Networks*, IEEE and INNS, Singapore.

[58] C.A.Anderson, K.F.Jones, M.Parker and J.Ryan (1993) Path assignment for call routing: an application of tabu search. *Annals of Ops.Res.*, **41**, (to appear).

[59] F.Glover and M.Laguna (1992) Bandwidth packing: a tabu search approach. *Man.Sci.*, (to appear).

[60] P.Hansen, B.Jaumard and M.Poggi di Aragao (1992) Mixed integer column generation algorithms and the probabilistic maximum satisfiability problem. *Proc. of the 2nd Integer Programming and Combinatorial Optimization Conference*, Carnegie Mellon.

[61] F.Glover and C.McMillan (1986) The general employee scheduling problem: an integration of management science and artificial intelligence. *Computers & Ops.Res.*, **15**, 563-593.

[62] J.Bovet, C.Constantin and D.de Werra (1992) A convoy scheduling problem. *Discrete Appl.Math.*, (to appear).

[63] M.Laguna and J.L.Gonzalez-Velarde (1991) A search heuristic for just-in-time scheduling in parallel machines. *J. of Intelligent Manufacturing*, **2**, 253-260.

[64] J.W.Barnes and M.Laguna (1992) Solving the multiple-machine weighted flow time problem using tabu search. *IIE Transactions*, (to appear).

[65] J.Skorin-Kapov (1990) Tabu search applied to the quadratic assignment problem. *ORSA J. on Computing*, **2**, 33-45.

[66] J.A.Bland and G.P.Dawson (1991) Tabu search and design optimization. *Computer-Aided Design*, **23**, 195-202.

[67] S.Oliveira and G.Stroud (1989) A parallel version of tabu search and the path assignment problem. *Heuristics for Combinatorial Optimization*, **4**, 1-24.

[68] F.Glover, C.McMillan and B.Novick (1985) Interactive decision software and computer graphics for architectural and space planning. *Annals of Ops.Res.*, **5**, 557-573.

[69] A.Hertz and D.de Werra (1987) Using tabu search techniques for graph coloring. *Computing*, **29**, 345-351.

[70] A.Hertz, B.Jaumard and M.Poggi di Aragao (1992) Topology of local optima for the k-coloring problem. *Discrete Appl.Math.*, (to appear).

[71] C.Friden, A.Hertz and D.de Werra (1989) Stabulus: a technique for finding stable sets in large graphs with tabu search. *Computing*, **42**, 35-44.

[72] M.Sun and P.G.McKeown (1993) Tabu search applied to the general fixed charge problem. *Annals of Ops.Res.*, **41**, (to appear).

Chapter 4

Genetic Algorithms

Colin R Reeves

4.1 Introduction

In some respects, to call the genetic algorithm (GA) paradigm 'modern' might seem to be stretching the truth, since the first developments in this field took place 30 years ago. However, most early applications were in the realm of Artificial Intelligence—game-playing and pattern recognition for instance. Some research was carried out on function optimization, but it is only recently that applications to problems of interest to the Operational Research community—such as the types of combinatorial optimization problem discussed in this book—have become more widely reported.

Section 4.2 will expound the basic concepts of GAs and explain why they work, while section 4.3 illustrates the methodology by a simple example. Extensions and modifications to the simple GA will be discussed in section 4.4, and some areas of application will be covered in section 4.5. While some of the theoretical results which justify the use of GAs will be discussed, in particular the fundamental *schema theorem*, this is not meant to be a theoretical text, and for a fuller account, the reader is referred to the books by Holland [1] and Goldberg [2]. Finally, although allusions will be made to some of the work carried out in other contexts, most of the discussion of applications will relate to the way GAs have been used for combinatorial optimization.

4.2 Basic Concepts

From an Operational Research perspective, the idea of a genetic algorithm can be understood as the intelligent exploitation of a random search. Viewed in this light, there are some interesting historical precedents in the context of combinatorial optimization.

Previously, there have been several attempts to solve problems by exploiting information obtained from randomly sampling the solution space. For example, the heuristic of Roberts and Flores [3] for the travelling salesman problem worked by initially generating many possible tours which represent a (biased) sample from the set of all tours. Links which appear rarely are eliminated, while ones which appear very often are committed, and the heuristic searches over a percentage of all the other links in order to find the best tour available. A similar idea was used by Nugent *et al.* [4] for solving the quadratic assignment problem under the name of *biased sampling.* In a different but related field, there have also been developments by statisticians of methods of experimental design—for example the work on response surface techniques described by Box and Draper [5].

However, while these methods do some of the same things as a genetic algorithm, the GA approach is far more flexible and provides a general framework for a wide variety of problems.

Genetic algorithms were developed initially by Holland and his associates at the University of Michigan in the 1960s and 1970s, and the first full, systematic (and mainly theoretical) treatment was contained in Holland's book *Adaptation in Natural and Artificial Systems* [1] published in 1975. Goldberg [2] gives an interesting survey of some of the practical work carried out in this era. Among these early applications of GAs were those developed by Bagley [6] for a game-playing program, by Rosenberg [7] in simulating biological processes, and by Cavicchio [8] for solving pattern-recognition problems.

The name *genetic algorithm* originates from the analogy between the representation of a complex structure by means of a vector of components, and the idea, familiar to biologists, of the genetic structure of a chromosome. In selective breeding of plants or animals, for example, offspring are sought which have certain desirable characteristics—characteristics which are determined at the genetic level by the way the parents' chromosomes combine. In a similar way, in seeking better solutions to complex problems, we often intuitively combine pieces of existing solutions. The parallel is not exact, but it was sufficiently persuasive for Holland to propose the problem-solving methodology which is described below.

In many applications, the component vector, or *chromosome*, is simply a string of 0s and 1s. Goldberg [2] suggests that there are significant advantages if the chromosome can be so structured, although a more recent argument by Antonisse [9] casts doubt on this. Nevertheless, much of the theoretical development is easier to understand if it is thought of in this way. Several *genetic operators* have been identified for manipulating these chromosomes, the most commonly used ones being *crossover* (an exchange of sections of the parents' chromosomes), and *mutation* (a random modification of the chromosome).

In this chapter we shall also refer to chromosomes as *strings*, *vectors*, or even *solutions*, depending on the context. Continuing the genetic analogy, variables are often called *genes*, the possible values of a variable *alleles*, and the position of a variable in a string is called its *locus*. In simple cases, the locus of a variable/gene is usually irrelevant, but as we shall see later, in more complex problems it becomes important. A further distinction is drawn, in genetics, between the chromosome (or a collection of chromosomes) as *genotype*, meaning the actual structure (or structures), and the *phenotype*—the physical expression of the structure (or structures). In terms of a GA, we may interpret the genotype as the coded string which is processed by the algorithm, while the decoded set of parameters represents the phenotype.

In the context of most obvious relevance to OR, that of finding the optimal solution to a large combinatorial problem, a genetic algorithm works by maintaining a population of M chromosomes—potential parents—whose *fitness values* have been calculated. Each chromosome encodes a solution to the problem, and its fitness value is related to the value of the objective function for that solution. In Holland's original GA, one parent is selected on a fitness basis (the better the fitness value, the higher the chance of it being chosen), while the other parent is chosen randomly. They are then 'mated' by choosing a crossover point X at random, the offspring consisting of the pre-X section from one parent followed by the post-X section of the other.

For example, suppose we have parents P1 and P2 as follows, with crossover point X; then the offspring will be the pair O1 and O2:

```
P1   1 0 1 0 0 1 0       01   1 0 1 1 0 0 1
         X
P2   0 1 1 1 0 0 1       02   0 1 1 0 0 1 0
```

One of the existing population is chosen at random, and replaced by one of the offspring. This *reproductive plan* is repeated as many times as is desired. In another version of this procedure, parents are chosen in strict proportion to their fitness values (rather than probabilistically), and the whole population is changed *en bloc* after every set of M trials, rather than incrementally.

Many other variations on this theme are possible: both parents can be selected on a fitness basis, mutation can be applied, more than one crossover point could be used, and so on, and several more sophisticated operators have been developed. However, in many cases, the simple GA involving just crossover and mutation has proved to be quite powerful enough.

In addition to these methodological decisions, there are various parametric choices, and it is also quite possible that the way the initial population is chosen will have a significant impact on the results. In fact, there is plenty of empirical evidence that Holland's original algorithm can be improved upon, although rigorous theoretical analysis of the modifications has not always been carried out. Before proceeding to discuss some of these modifications, we give a brief overview of the fundamental theoretical results which underpin the application of GAs.

4.2.1 Intrinsic parallelism and the schema theorem

The underlying concept Holland used to develop a theoretical analysis of his GA was that of a *schema*. (The word comes from the past tense of the Greek verb $\epsilon\chi\omega$ (*echo*, to have), whence it came to mean shape or form; its plural is *schemata*). If we think of each chromosome as a vector of 0s and 1s, then for instance, the two vectors

```
1 0 1 0 0 0 1
1 1 1 1 0 1 0
```

are both examples of the schema

$$* \; * \; 1 \; * \; 0 \; * \; *$$

where the $*$ symbol is a 'wild card', i.e. it can be replaced by a 0 or a 1. Schemata can be thought of as defining subsets of similar chromosomes, or as *hyperplanes* in n-dimensional space. Another interpretation is in terms of periodic 'functions' of different 'frequencies'—for example, as illustrated in Figure 4.1 the schema $* \cdots * 1$ represents the odd integers in the case where the chromosome (genotype) encodes integer values (phenotype). This concept has in fact been important in some of the most recent advances in the theoretical analysis of GAs.

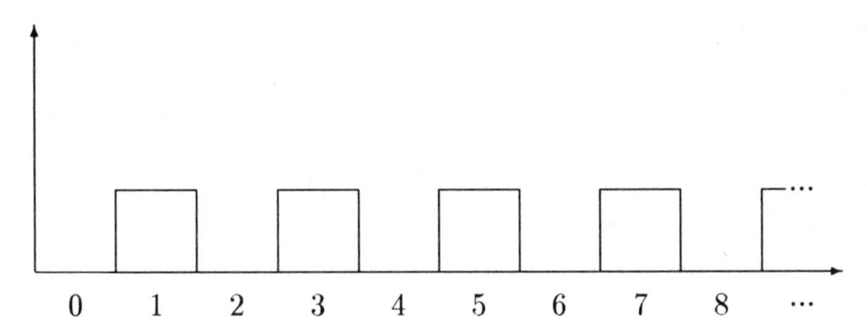

Figure 4.1: Possible integer values of the schema $* \cdots * 1$

Returning to the example, it is clear that the chromosomes are also instances of several other schemata. Some schemata contain both chromosomes, while others have only one of them as a member. In general, if the string has length n, each chromosome is an instance of 2^n distinct schemata, since at each locus it can take either its actual value or a $*$. As a consequence of this, each time we evaluate the fitness of a given chromosome, we are gathering information about the average fitness of each of the schemata of which it is an instance. In theory, a population could contain $M2^n$ schemata, but of course in practice there will be some overlapping so that not all schemata will be equally represented, and some may have no representatives at all. (In fact we want them to have unequal representation: we want to focus attention on the ones which are fitter.) It is in this sense—that we are testing a large number of possibilities by means of a few trials—that a GA possesses the property which Holland calls *intrinsic parallelism.* (Some other authors use the alternative term *implicit* parallelism.)

The question is of course, how many schemata are being processed? Under certain plausible assumptions, it can be shown that processing a population of size M in one generation in fact processes $\mathcal{O}(M^3)$ schemata. This result will be explained in greater detail later.

However, it is clearly impossible to store the average fitness values explicitly, nor is it obvious how to exploit such information even if we could. Holland's second important insight was to show that by the application of genetic operators such as crossover and mutation, each schema represented in the current population will increase or decrease in line with its relative fitness, independently of what happens to other schemata.

In mathematical terms, he proved what has become known as the schema theorem, in which the notions of the *length* and *order* of a schema are used. These are respectively the *distance* between the first and last defined (i.e. non-$*$) positions on the schema, and the *number* of defined positions. Thus, the schema described above has length 2 and order 2. Another important parameter is the *fitness ratio* which is defined as the ratio of the average fitness of a schema to the average fitness of the population. As this is so fundamental to the understanding of why a GA works, this result will be covered in a little more depth. We approach this theorem by proving the following results:

Lemma 4.1 *Under a reproductive plan in which a parent is selected in proportion to its fitness, the expected number of instances of a schema S at time $t+1$ is given by*

$$E(S,t+1) = f(S,t)N(S,t)$$

where $f(S,t)$ is the fitness ratio for schema S, and $N(S,t)$ is the number of instances of S, at time t.

Proof The result follows immediately from the definition of the reproductive plan. □

Lemma 4.2 *If crossover is applied at time t with probability P_c to a schema S of length $l(S)$, then the probability that S will be represented in the population at time $t+1$ is bounded below by*

$$P[S,t+1] \geq 1 - \frac{P_c\, l(S)}{n-1}(1 - P[S,t])$$

where n is the length of the chromosome.

Proof The probability that schema S will be destroyed by the crossover operator is easily seen to be the product of the probability that crossover is applied, P_c, and the probability that the crossover site lies within the length of the schema, $l(S)/(n-1)$.

If the other parent is itself an instance of S, then S will survive regardless of the crossover site; hence this must be multiplied by $(1 - P[S,t])$—the probability that the second parent is an instance of a different schema, assuming it is chosen randomly using a uniform distribution—to give the probability of destruction.

Hence the probability of non-destruction is

$$1 - \frac{P_c\, l(S)}{n-1}(1 - P[S,t])$$

Finally, this is clearly only a lower bound, since it is also possible for S to appear at time $t+1$ as a result of crossover of two different schemata S' and S''. □

Lemma 4.3 *If mutation is applied at time t with probability P_m to a schema S of order $k(S)$, then the probability that S will be represented in the population at time $t+1$ is bounded below by*

$$P[S,t+1] \geq 1 - P_m k(S)$$

Proof From the definition of mutation, the probability that mutation fails to destroy S is given by

$$(1 - P_m)^{k(S)} \geq 1 - P_m k(S)$$

□

Combining these results together we obtain the following theorem:

Theorem 4.1 (The Schema Theorem) *Using a reproductive plan as defined above in which the probabilities of crossover and mutation are P_c and P_m respectively, and schema S of order $k(S)$, length $l(S)$ has fitness ratio $f(S,t)$ at time t, then the expected number of representatives of schema S at time $t+1$ is given by*

$$E(S,t+1) \geq \{1 - \frac{P_c\, l(S)}{n-1}(1 - P[S,t]) - P_m k(S)\} f(S,t) N(S,t)$$

(It should be noted that the statement of this theorem is slightly different from that sometimes found. In particular, the left-hand-side is explicitly stated in terms of *expectation*, a fact which is often insufficiently stressed; also the effect of mutation is included.)

Consideration of this expression shows that the representation of S in the population will increase on the average provided that

$$f(S,t) \geq 1 + \frac{l(S)}{n-1} + P_m k(S)$$

Thus, short low-order schemata will increase their representation provided their fitness ratio is slightly more than 1; longer high-order schemata have to work harder.

Thus the ideal situations for a GA are those where short, low-order schemata combine with each other to form better and better solutions. The assumption that this will work is called by Goldberg [2] the *building-block hypothesis.* Empirical evidence is strong that this is a reasonable assumption in many problems.

The $\mathcal{O}(M^3)$ Estimate

We now return to the question of how many schemata are being processed by a GA. To answer this in general is a formidable task, but we can simplify by first restricting the question to those schemata which have a probability $1-\epsilon$ of surviving. (The value ϵ can be thought of as a *transcription error*, and is of course a function of schema fitness, length and order, as given by the schema theorem.) This further implies that we need only consider schemata of *length* $\leq 2L$, say, where L is a function of ϵ. Finally, we will count schemata of *order* L only.

In order to estimate the number of schemata satisfying these conditions, we consider a 'window' of $2L$ contiguous loci counting from the leftmost position. There are $\binom{2L}{L}$ possible choices of defining loci for an order L schema in this window, each of which would give rise to 2^L actual schemata. (For example, $f * f * \cdots *$, where f is a fixed value of 0 or 1, represents 4 schemata corresponding to the 4 ways of choosing the fixed values.) We now successively move this window 1 step to the right, observing that each time there will be $\binom{2L-1}{L-1}$ choices of defining loci which have not been counted in the previous window. Altogether, we can move this window 1 step to the right $(n - 2L)$ times, so there are

$$\{\binom{2L}{L} + (n - 2L)\binom{2L-1}{L-1}\}2^L = (n - 2L + 2)\binom{2L}{L}2^{L-1}$$

possible schemata per string $\approx 2^{3L-1}/\sqrt{\pi L}$, using Stirling's approximation.

Of course, in a given string, all 2^L schemata are not represented for a particular choice of defining loci, so lastly, we need to estimate how many schemata meeting the prescribed conditions will occur in a population of size M. On the assumption that the members of the population are chosen using uniform random sampling, a population of size $M = \nu 2^L$ (for ν a small integer value) would be *expected* to have ν instances of each of these order L schemata, so that with this definition of M, we can estimate the number of schemata of length $2L$ or less, and order L, as $\mathcal{O}(M^3)$. In fact, this is probably an underestimate, since there are many other schemata we have not counted at all. But for the purpose of demonstrating the existence of intrinsic parallelism, it is sufficient.

4.2.2 Recent developments

Theoretical developments have continued since the original work by Holland. Schaffer [11] has analysed the effect on the schema theorem of either selecting both parents uniformly, or both on a fitness basis. The latter case in particular can be shown to improve the chances of good but longer schemata. Bridges and Goldberg [13] extended the schema theorem from a lower bound to calculate an exact expression for the expected number of a given schema in the population.

More recently, the use of Walsh functions has led to significant advances in the analysis of GAs. Walsh functions can be thought of as discrete analogues of trigonometrical functions; they were originally developed by mathematicians and applied by engineers to the field of signal processing. Beauchamp [15], for instance, gives a thorough introduction to the subject. Their application to GAs springs from the 'periodic function' interpretation of schemata. Unfortunately, to do full justice to these important developments would require a much longer chapter than this one, but it is possible briefly to describe some of this work.

Walsh function analysis of GAs was first introduced by Bethke [16], and has been greatly extended in a series of important papers by Goldberg [17, 18, 19] and his associates. Firstly, Goldberg and Rudnick [19] have derived an expression for the variance of schema fitness. The schema theorem is a statement about expectations, and it is often observed in practice that the fittest schemata have a substantial probability of failing to survive, merely because of the variation in the bits of the string whose values are not fixed. (Goldberg and Rudnick term this variation 'collateral noise'.) In order to prevent this, larger populations are needed; their results enable an estimate to be made of the appropriate population size.

In another application of Walsh functions to the analysis of GAs, Goldberg [18] has introduced the idea of *deceptive problems*. While the building-block hypothesis has been found reliable for many problems, it is not hard to imagine cases where it fails. An intuitive idea of such problems is that they possess optimum points which are isolated, so that adjacent points lead the GA away from it. (Walsh function analysis helps to define this idea more rigorously.) Another recent paper by Whitley [20] has analysed the concept of deception in greater depth, and he shows that the implicit parallelism property fails for precisely these types of problems.

Deceptive problems have been shown to yield to another type of GA—a 'messy' GA (mGA) [21, 22]. Again, a full account of this concept is beyond the scope of this chapter, and what follows is merely an outline. The central idea is to start with a very large population of small sub-units each representing specified values for a few positions of the string. These are evaluated for fitness by means of a scheme called 'competitive templates', whereby the unspecified positions of a string are given the values which they would have in a local optimum generated by some other method. (In effect, at the initial level at least, the mGA is processing schemata directly, rather than operating on strings.) The small units are then selected for recombination as in a conventional GA, and mated using a 'messy' version of the usual crossover operator. In order to create larger units, the algorithm may actually have to make a choice between alternative specifications for the same position—this accounts for some of the modifications to ordinary crossover—but fairly rapidly the population is refined into longer strings representing better and better solutions. Experimental results for this approach are impressive, particularly for those deceptive problems which defeat a conventional GA; furthermore, Goldberg *et al.* [22] were able to establish excellent theoretical convergence properties for the mGA.

Another interesting recent development is Radcliffe's [23] concept of a *forma* (Latin for form or shape; its plural is *formae*). Whereas a schema defines subsets of chromosomes which are similar in the specific sense that they have the same alleles at some specified loci, a forma defines subsets of chromosomes which are similar in some possibly more general way. For example, in the case of the travelling salesman problem, the defining characteristic of a forma might be that a chromosome possesses a particular edge. Traditional schema analysis is able to explain *a posteriori* why the simple crossover 'works'. With Radcliffe's approach, it is possible to invert the process, and, *a priori*, design operators to have desirable properties. Two such properties are those he calls *respect* and *proper assortment.* Informally, respect implies that where parents share a particular characteristic, their offspring should always inherit this characteristic, while proper assortment implies that where parents have different characteristics, it should at least be possible that their offspring will contain both characteristics. (Proper assortment can thus be interpreted as a more systematic statement of the building-block hypothesis.) These properties are not always compatible, and the operators which can be built from their requirements would seem often to be computationally more expensive than simple crossover and mutation.

In fact, the simple GA has been found to work effectively in many situations, and the extra complications of the messy or forma-related versions have yet to be used in many applications. Accordingly, while these ideas may prove significant in the future, they are not a major concern of this book, and we return here to the further consideration of the conventional genetic algorithm.

4.3 A Simple Example

To make the action of a GA more transparent, it will be described in the context of maximizing a simple function

$$f(x) = x^3 - 60x^2 + 900x + 100$$

While this is not a *combinatorial* problem, it will illustrate the basic features of a GA which are common to all optimization problems. (In the context of this book, it would have been more convenient if we could have used a combinatorial problem as an illustration; however, as we shall see, such problems require the inclusion of constraints and/or more complicated encodings which would draw attention away from these basic features.)

To keep things simple, we will assume that we are searching for the maximum which is known to be an integer in the range $[0, 31]$. In fact, it is easily checked that $f(x)$ has a maximum at $x = 10$ and a minimum at $x = 30$. We can therefore easily represent x (the phenotype) in binary format as a 5-bit string of 0s and 1s (the genotype).

Suppose now we generate 5 random strings with the results as shown below.

No.	String	x	$f(x)$	$P[select]$
1	10011	19	2399	0.206
2	00101	5	3225	0.277
3	11010	26	516	0.044
4	10101	21	1801	0.155
5	01110	14	3684	0.317
Average fitness			2325	

Treating the $f(x)$ values simply as fitness values, the probability of selecting each string as Parent 1 is in direct proportion to its value, and is given in the $P[select]$ column. By drawing uniform random numbers between 0 and 1, we select Parent 1 using this distribution, and Parent 2 using a discrete uniform distribution. Again, by drawing uniform random numbers, we proceed to carry out crossover (with

$P_c = 1$) at a randomly determined site, and apply mutation (with $P_m = 0.02$). The results of a typical experiment were as follows:

Step	Parent 1	Parent 2	Crossover point	Mutation?	Offspring String	$f(x)$
1	1	2	4	No	10011	2399
2	5	3	2	No	01010	4100
3	5	2	3	No	01101	3857
4	4	2	1	No	10101	1801
5	2	5	4	No	00100	2804
	Average fitness					2992

As this is a very small population, we could not expect it to display every facet of a GA's operation. For instance, the relative frequencies of selection do not match the *P*[*select*] values very well, and no mutations took place. However, we can see that Chromosome 5, the best in the original population, has been selected more often as Parent 1 than any of the others, and Chromosome 3, the worst, has not been selected at all as Parent 1. Further, the average fitness has risen quite substantially, and we have in fact found the maximum at step 2.

In terms of schemata, there is not much to say from this small example, but at least we can see that the schema $1 * * * *$ (which can never give the maximum) has been reduced from 3 copies to 2 in one generation; conversely schema $0 * * * *$ (defining a hyperplane which *does* contain the maximum) has increased from 2 copies to 3. In the interests of brevity, further generations have not been displayed, but the example can easily be extended further by the reader.

Even in this small example, there are many variations we could use.

- We generated a whole new population to replace the original one *en bloc*, but we could have inserted each new chromosome as it was formed in place of a randomly chosen member of the current population.

- Each crossover could produce two offspring, but we only used one.

- We used stochastic selection with replacement which is a highly variable procedure; sampling without replacement might get closer to the expected frequencies.

- We used the *raw* $f(x)$ values to define fitness, but some sort of scaling procedure might well give superior discrimination between good and bad chromosomes.

4.4 Extensions and Modifications

As we have just seen, the basic scheme suggests a wide variety of different ways of implementing a GA, and exploring these has produced a large literature. To some extent the development of GA practice has been rather chaotic, and the terminology has not had time to settle down. It would also appear that in a number of cases similar ideas have been 'discovered' independently by more than one person. It is hoped that one benefit of the discussion that follows will be that readers will be saved from 're-inventing the wheel' in implementing their own GAs. In many of these cases theoretical work has yet to be done and the justification for using them is empirical. The variety of developments is large, and it would easily be possible to devote a whole book to the subject. This chapter will simply try to cover the most important and effective modifications; for convenience they will be categorized into those modifications that relate to the population, those that relate to coding and representation of chromosomes, and those that relate to the genetic operators themselves. (While some of these variations relate to the more general optimization situation, the main focus will be on extensions that have some relevance to combinatorial problems in particular.) Finally, there will be some discussion of *hybridization* of GAs, and a brief reference to some *parallel* implementations.

Before embarking on this survey, it may be helpful to draw attention to the way GA researchers usually evaluate GA performance, as it differs somewhat from the methods used in more traditional work on combinatorial optimization. Following De Jong [24], three measures of performance are generally used: *on-line, off-line* and *best-so-far*. The definition of *best-so-far* is obvious; indeed, it is the usual measure of performance when evaluating any iterative heuristic technique, but the other two need further explanation. If the objective function corresponding to the chromosome generated at time t (iteration t) is $v(t)$, and the best value found up to (and including) time t is $v^*(t)$, then after T iterations (function evaluations)

$$v^{online}(T) = \frac{\sum_{t=1}^{T} v(t)}{T}$$

$$v^{offline}(T) = \frac{\sum_{t=1}^{T} v^*(t)}{T}$$

On-line performance measures the average value of all strings generated, whereas the *off-line* measure relates more to the performance of the GA in converging to an optimum. The latter puts stress on the attainment of (possibly isolated) good solutions, the former on the GA's ability to find an area of the search space where performance is good in general.

The reasons for making these distinctions do not always apply, but as much of the GA literature makes use of them, it is helpful to mention them. Naturally, in specific circumstances, other measures have also been used: for example, the number of iterations to reach the global optimum (where it is known), or to reach a given standard (where it is not).

4.4.1 Population-related factors

Population size

One of the most obvious questions relating to GA performance is how it is influenced by population size. In principle, it is clear that small populations run the risk of seriously under-covering the solution space, while large populations incur severe computational penalties. Some theoretical work has been reported on this trade-off by Goldberg [25]. This indicates that the optimal size for binary-coded strings grows exponentially with the length of the string n.

However, this would imply extremely large populations in most real-world problems, and the *practical* performance of the GA would be quite uncompetitive with other methods such as simulated annealing and tabu search. Fortunately, empirical results from many authors suggest that population sizes as small as 30 are quite adequate in many cases, while some experimental work by Alander [26] suggests that a value between n and $2n$ is optimal for the problem type considered. Goldberg's later results, based on a different argument (as reported above [19]), go some way to supporting the use of populations rather smaller than his earlier work suggested.

Seeding

A subject that has attracted little attention is the question of the selection of the initial population. Usually, this is simply chosen by generating random strings of 0s and 1s. Some reports [27, 28] have found that 'seeding' a population with a high-quality solution, obtained from another heuristic technique, can help a GA find better solutions rather more quickly than it can from a random start. However, there is a possible disadvantage in that the chance of premature convergence may be increased.

Selection mechanisms

In Holland's original GA, parents were selected by means of a stochastic procedure from the population, and a complete new population of offspring was generated which then replaced their parents *en bloc*. In another version, he suggested that each offspring should replace a randomly chosen member of the current population as it was generated. De Jong [24] introduced the idea of a generation gap to allow parents and offspring a more controlled amount of overlap. A proportion G was selected for reproduction, and their offspring replaced randomly selected existing population members. (Holland's original GAs correspond to $G = 1$ and $G = M^{-1}$ respectively.)

Studies seem to indicate that GAs perform better when populations do not overlap, but the $G = M^{-1}$ case (*incremental* or *steady-state* replacement) in particular has been used with success, and it has certain advantages in terms of ease of implementation. It also seems easier, as Davis [29] among others has pointed out, to prevent the ocurrence of duplicates by using incremental replacement. Duplication is unhelpful: firstly, it wastes resources on evaluating the same fitness function twice over (or else resources must be used up in detecting the existence of duplicates); secondly, it distorts the selection process, by giving extra chances to the duplicate chromosomes to reproduce. Computational experience with the GENITOR algorithm developed by Whitley [30] supports the argument that incremental replacement is more efficient than replacement *en bloc.*

An obvious defect (from an optimization viewpoint) with the simple GA is that there is no guarantee that the best member of a population will survive into the next generation. One way of dealing with this is De Jong's *élitist* model, in which the best member of the current population is forced to be a member of the next. Another approach was taken by Ackley [31], who coined the term *termination with prejudice* for his suggestion. This was to use an incremental replacement approach where at each step the new chromosome replaced one randomly selected from those which currently have a *below-average* fitness.

The selection of parents is subject to sampling errors, which lead to a sometimes serious difference between the actual and expected number of times a chromosome is used. De Jong [24] suggested using an *expected value* model, where chromosomes are forced to become parents more or less in line with their expected frequencies as predicted by their fitness values, by following a policy of random sampling without replacement. This approach consistently out-performed the conventional one of sampling *with* replacement. Further refinements [32, 33, 34] have produced other ways of selecting parents in order to improve performance.

Many other subtleties in selection mechanisms have been developed; there is one more which is worth a specific mention—the idea of *speciation*. Real populations do not favour indiscriminate mating, for the result (even if it is biologically possible) is usually sterile. In a similar way, if we try to optimize a multi-modal function by means of a GA, the result from crossover of two chromosomes which are close to *different* optima may be much worse than either. The answer in nature is the existence of *species*, and there have been several attempts [24, 35, 36] to replicate this behaviour in the context of GAs. The method of Goldberg and Richardson [35] was to define a *sharing function* over the population which is used to modify the fitness of each chromosome. The function could take many forms, but a simple linear function

$$\begin{aligned} h(d) &= 1 - d/D, \; d < D \\ h(d) &= 0, \; d \geq D \end{aligned}$$

is effective. Here d is the *distance* between two chromosomes, and D is a parameter. The idea is that the sharing function is evaluated for each pair of chromosomes in the population, and then the sum

$$\sigma_j = \sum_{i \neq j} h(d_i)$$

is computed for each chromosome j. Finally the fitness of chromosome j is adjusted by dividing by σ_j; the adjusted fitness values are then used in place of the original values. The effect of this is that chromosomes which are close will have their fitness devalued relative to those which are fairly isolated. The only question is how to measure distance. The obvious measure on the chromosomes is Hamming distance (for binary strings); however, there is a well-known problem with binary coding—numbers which are actually close can be mapped to binary representations which are far apart and *vice-versa*. It is therefore preferable to define distances in terms of the actual numerical values of the parameters which are encoded in the binary string—i.e. to use phenotype distances rather than genotype distances.

Fitness calculation

So far, we have assumed that a fitness measure is available, without discussing how it might be obtained. A naive choice for an optimization problem is simply to use the value of the objective function associated with each chromosome, but this is rarely a good idea.

A phenomenon that has often been observed is that as the algorithm proceeds, the population converges to a set of very similar chromosomes between which it is hard to discriminate if we use a naive fitness measure. A slightly different problem often arises at the start of the process, when there are many poor chromosomes and just one or two outstanding ones. Here the naive fitness measure may lead to a rapid takeover by the latter, and premature convergence to a poor local optimum.

Both of these problems can be mitigated by using a scaling procedure. The requirement is to limit competition early on, but to stimulate it later. A simple means of doing this, if the objective function value is v, and the required fitness value is f, is to use the transformation

$$f = av + b$$

where the values a, b are obtained from the conditions:

$$f_{mean} = v_{mean}$$

$$f_{max} = \mu v_{mean}$$

The parameter μ is used to ensure that the fittest member of the population will be chosen on average μ times. A difficulty can arise if the scaled fitness values become negative, but circumventing this is fairly straightforward. Goldberg [2] discusses some of these possibilities.

Another approach is to ignore the actual objective function values and use a ranking procedure instead. Baker [37] considered schemes of this type, while Reeves [27] also used this idea, with some success, in building a GA for flowshop sequencing. Here, potential parents were selected using the probability distribution

$$p([k]) = \frac{2k}{M(M+1)}$$

where $[k]$ is the kth chromosome when they are ranked in ascending order. This distribution seems to give a sensible selective pressure, in that the best (chromosome $[M]$) will have a chance of $2/(M+1)$ of being selected, roughly twice that of the median, whose chance of selection is $1/M$. The paper by Whitley [30], to which reference was made above, generalizes this ranking method, and also contains a helpful discussion in which it is argued that ranking should be preferred to scaling. The main argument is that the key to good GA performance is to maintain an adequate selective pressure by means of an appropriate *relative* fitness measure. Ranking provides a means of doing this consistently regardless of the 'actual' fitness measure, which as he points out, is often an arbitrary value in any case.

An alternative which neatly combines the idea of ranking with the selection mechanism is to use *tournament selection.* In this procedure, a chromosome list is defined by randomly permuting their index numbers $1, \ldots, M$. Successive groups of T chromosomes are then taken from this list and compared, the best one being chosen as a parent. When the M chromosomes are exhausted, another random permutation is generated. The whole procedure is repeated until M

parents have been chosen in this way. Each parent is then mated with another chosen purely at random. It is easy to see that by this means the best chromosome will be chosen T times in any series of M tournaments, the worst not at all, and the median on average once. In the case $T = 2$, the effect is very similar to the ranking scheme described above, but without having to do any 'book-keeping' in maintaining an ordered list.

4.4.2 Modified operators

In many applications, the simple crossover operator has proved extremely effective; the addition of mutation merely helps to preserve a reasonable level of population diversity—an insurance policy which enables the process to escape from sub-optimal regions of the solution space. Usually, the procedure appears to be fairly tolerant of a range of mutation probabilities, although setting the rate too high could of course prevent crossover from doing its work properly. However, despite the general success with just these two operators, there are cases where more advanced operators have been found useful.

'String-of-change' crossover

A frequent observation when applying a GA in practice is that, particularly in the later stages of a run, the parent chromosomes converge to such an extent that crossover has little effect. For instance, the two strings

```
1 1 0 1 0 0 1
1 1 0 0 0 1 0
```

will fail to generate a different string if the crossover point is any of the first three positions. Booker [39] and Fairley [40] have independently suggested that before applying crossover, we should examine the selected parents to find the crossover points which would produce offspring which differ from the parents. Booker found some improvement by restricting crossover points in this way, but overall his results were somewhat inconclusive. Fairley implemented this idea by his 'string-of-change' crossover (Booker used the term 'reduced surrogates'), which entails computing an 'exclusive-OR' (XOR) between the parents. Only positions between the outermost 1s of the XOR string will be considered as crossover points. Thus in the example above, the XOR string is

```
0 0 0 1 0 1 1
```

so that, as previously stated, only the last 3 crossover positions will give rise to a different string. Fairley investigated this operator in the context of solving knapsack problems, and found significantly improved performance compared with the standard crossover operator.

Multi-point crossover

There seems no reason why the choice of crossover point should be restricted to a single position, and Booker [39] for example, has empirically observed that increasing the number of crossover points to two has improved the performance of a GA. The most thorough investigation of multi-point crossovers is that by Eshelman *et al.* [41], who examined the biasing effect of traditional one-point crossover, and considered a range of alternatives. Their central argument is that two sources of bias exist to be exploited in a genetic algorithm: *positional* bias, and *distributional* bias. Simple crossover has considerable positional bias, in that it relies on the building-block hypothesis, and if this is invalid, the bias may be *against* the production of good solutions.

Simple crossover has no distributional bias, in that the crossover point is chosen randomly using the uniform distribution. But this lack of bias is not necessarily a good thing, as it limits the exchange of information between the parents. In [41], the possibilities of changing these biases, in particular by using multi-point crossover, were investigated and empirical evidence strongly supported the suspicion that one-point crossover is not the best option. In fact, the evidence was somewhat ambiguous, but seemed to point to an 8-point crossover operator as the best overall, in terms of the number of function evaluations needed to reach the global optimum. This work relates to the application of GAs to solving functional optimization problems, rather than combinatorial problems, and while it appears unlikely that results would be dramatically different, there is little reported on multi-point crossovers in combinatorial optimization.

Generalized crossover

Several reports have suggested ways of generalizing the simple crossover operator. The latter can itself be represented as a binary string: for example, the operator representation

```
1 1 1 0 0 0 0
```

may be taken to mean that the first 3 elements of a 7-bit chromosome are taken from the first parent, and the last 4 from the second.

This can clearly be generalized very easily by allowing the pattern of 0s and 1s to be generated stochastically using a Bernoulli distribution; thus a template such as

```
1 0 1 0 0 0 1
```

implies that the 1st, 3rd and 7th elements are taken from the first parent, the other elements from the second.

Syswerda [42] investigated this 'uniform crossover' and discussed its advantages and disadvantages over simple crossover. The main advantage is that 'building-blocks' (in Goldberg's phrase) no longer have to be encoded compactly (i.e. as short schemata) in order to survive. Uniform crossover is obviously completely indifferent to the length of a schema: all schemata of a given order have the same chance of being disrupted. Of course in some problems this may not be an advantage at all!

Sirag and Weisser [43] take yet another route, by modifying the basic genetic operators in the spirit of simulated annealing. Thus, for example, the crossover operator is modified by defining a *threshold energy* θ_c which influences the way in which individual bits are chosen. Briefly, as the offspring chromosome is generated, there is a presumption in favour of taking bit $i+1$ from the same parent as bit i. However, bit $i+1$ is taken from the other parent with probability

$$\exp(-\theta_c/T)$$

where T is a 'temperature' parameter, which is slowly decreased by an 'annealing schedule'. At high temperatures, this can be expected to behave rather like the generalized uniform operator; as temperatures moderate the number of 'switches' between parents decreases and it becomes more like the standard simple crossover, while at very low temperatures it just copies one of the parents. Choosing the 'best' values of θ_c and T may require some experimentation, as will the annealing schedule.

Recently, many of these extensions have received a theoretical treatment by De Jong and Spears. In their paper [44] they are able to characterize the amount of disruption introduced by a given crossover operator exactly. In the course of this work, they further generalize the uniform crossover operator: in Syswerda's original description it was implicitly assumed that the Bernoulli parameter was 0.5, but De Jong and Spears show that (as is intuitively expected) the amount of disruption can be reduced by choosing different values.

Inversion

Simple crossover implicitly assumes that there is no order relationship between adjacent genes. In naturally occurring chromosomes the case is often otherwise, and re-ordering operators may well be important. *Inversion* is one operator which takes account of order relationships. As the name implies, a section of the chromosome is 'cut out' and then re-inserted in the reverse order[1].

Thus the chromosome S1 below is transformed into S2 by inverting the section X-Y.

```
S1   1 0 1 0 1 1 0        S2  1 0 1 1 1 0 0
          X     Y
```

However, inversion has not often been found significantly useful; if used at all, it is best used in a similar rôle to that of mutation—it may allow the exploration of regions of the search space that crossover will not quickly reach, but in general it does not have the re-combinative power of crossover.

Dominance

Naturally occurring chromosomes are often *diploid*, that is, at each locus of the chromosome there is a *pair* of genes. However, only one of the pair is actually *expressed*, and this is called the *dominant* gene.

[1]There is some ambiguity in the literature regarding the concept of inversion: Holland's original description [1] is more complex than that adopted here, and is only effective if combined with a modified crossover operator.

The other is called *recessive*, and can only be expressed if paired with another recessive.

Several attempts have been made to find ways of copying and exploiting this behaviour in the realm of genetic algorithms. The practical benefit of diploid dominance is that a form of long-term memory can be preserved, whereas with *haploid* (single-stranded) chromosomes there is the danger of useful information disappearing altogether.

Goldberg and Smith [45] present an interesting application of dominance in an OR context. The problem they describe is a non-stationary Knapsack Problem with a single constraint whose right-hand-side periodically alternates between two different values. A haploid GA finds this difficult: a good solution for the first value may be hopeless for the second, and usually it fails to find a second solution before the alternative RHS becomes operative again. A diploid GA overcomes this by preserving the memory of good solutions in both situations, so that it is possible to adapt fairly quickly every time the RHS changes. However, in the main, diploidy has not been found to be important in applications of GAs.

Adaptive operator probabilities

Most GA implementations have assumed that the probability or rate of using a particular operator is fixed throughout. Mutation is applied with a low probability (typically less than 1%) at each locus of a chromosome which has already been generated by crossover applied with a high probability (often 100%) to two chromosomes. Thus crossover-plus-mutation is effectively a single operator, applied in accordance with probabilities fixed at the outset. Reeves [27] found it fruitful to modify this approach by allowing the mutation rate to change as the algorithm proceeds. The idea was to make the mutation rate inversely proportional to the population diversity, in order to prevent or at least alleviate the problem of premature convergence.

Booker [39] reported significant gains (in a function optimization context) from using an adaptive crossover rate: the rate was varied according to a characteristic called *percent involvement*. This is simply the percentage of the current population which is producing offspring—too small a value is associated with loss of diversity and premature convergence.

Davis [29] has suggested a different procedure where either crossover or mutation should be applied at each iteration but not both. Thus at each step the algorithm should choose between the operators on the basis of a probability distribution which he calls 'operator fitness'. Thus if crossover has a 'fitness' of 75% and mutation 25%, clearly crossover would be chosen 3 times as often as mutation.

In a further development of this idea, he suggested that the operator fitness values should themselves change as the algorithm proceeds. In particular, crossover is more important at the beginning when the population is diverse, but as the chromosomes start to converge it is important to increase the chance of finding different solutions, which is where mutation is most effective. It is possible to alter these values in a mechanical way, but Davis argues that it is better to allow the GA to find appropriate values by keeping track of the relative performance of crossover and mutation in producing better solutions, apportioning credit to the operators in proportion to their performance.

It is perhaps surprising that it is only recently that the question of whether there is an optimal mutation rate has excited much attention. De Jong [24] has been quoted as recommending that the bit-mutation rate should be n^{-1} where n is the string length, but although his work was some time ago, most reported applications have continued to use low rates which implicitly ignore the string length. Harvey [46] has applied some concepts from theoretical biology which suggest mutation rates of the same order of magnitude, while Schaffer *et al.* [47] have used experimental results to estimate the optimal rate as $\propto 1/M^{0.9318}n^{0.4535}$.

4.4.3 Chromosome coding and representation

Binary versus Gray coding

Although the 0-1 integer setting is the obvious one in which to use a GA, they have also been used successfully for problems requiring real values. Usually, this has entailed mapping real numbers onto a binary string to the desired degree of precision. Thus if a parameter which can take values between a and b is mapped to a string of length n, the precision is

$$\frac{a-b}{2^n-1}$$

However, there is a well-known practical problem with binary coding, in that values which are close in the original space may be far apart in the binary-mapped space, so that adjacent genotypes may have distant phenotypes, and *vice-versa.* For example, suppose the optimum occurs at a value 32 in numerical value, and we are using a 6-bit chromosome, so that the binary mapped value is 1 0 0 0 0 0. A good approximation to this value is 31, which maps to 0 1 1 1 1 1. Conversely, the binary value 0 0 0 0 0 0, only one bit different from the optimum, represents the numerical value 0, some distance away in the original space! This has led Caruana and Schaffer [48] to advocate the use of a Gray code mapping, but this introduces further problems in that there is no simple algorithm for decoding a Gray code. These authors have continued to champion the use of Gray coding [41], but most reported GA implementations still appear to use the simple binary coding. In the case of combinatorial optimization, the problem may not be so critical, as the bits may represent true 0-1 variables in any case, while in other situations a sequence representation (see below) is more appropriate.

Non-binary coding

As observed earlier, most GA implementations assume the original 0-1 structure to represent chromosomes. The argument in its favour is that it maximizes the number of schemata per bit in the string. However, this rests on a particular interpretation of the symbol $*$ in the specification of a schema as representing a set of strings in which the $*$ is replaced by *any* of the possible symbols. Antonisse [9] provides an alternative interpretation in which the $*$ represents *any subset* of the possible symbols. The implication is that where a binary alphabet is not a natural coding for a problem, it is perfectly reasonable to consider an alternative coding in a higher-cardinality alphabet. Indeed, it may actually be preferable even when a binary coding is fairly straightforward.

Recent empirical evidence bears this out. Reeves and Steele [49] have used chromosomes in which the alleles at each locus were defined using alphabets of varying cardinalities to encode parameters for configuring a neural network. In fact, the number of possible values at different positions on the chromosome varied from 2 to 40. In the context of function optimization, Davis [29] discusses the possibilities suggested by such encodings in more detail; for example, new versions of traditional operators such as 'averaging crossover' and 'creep mutation' operators can be invented. The concept of averaging is obvious; by 'creep' Davis means moving a component up or down relative to its current value by a small amount. Some analysis of operators in this context is presented by Radcliffe [23]. Of course, this has less relevance to *combinatorial* optimization, but there may be situations where such ideas could be exploited.

Sequence representation

Many problems of interest in OR can be most naturally represented as a permutation, the obvious one being the infamous Travelling Salesman Problem (TSP). Unfortunately, the ordinary crossover operator is no use in such situations, as can easily be seen from the following example.

```
P1   2 1 3 4 5 6 7     O1   2 1 3 2 5 7 6
          X
P2   4 3 1 2 5 7 6     O2   4 3 1 4 5 6 7
```

The inversion operator could be applied to *single* strings, but the power of a GA is in the recombining of *different* chromosomes. Several authors have therefore tried to define a crossover-like operator for such problems as the TSP.

Goldberg and Lingle [50] defined an operator called PMX (partially mapped crossover), which used *two* crossover points. The section between these points defines an interchange mapping. Thus, in the example above, PMX might proceed as follows:

```
P1   2 1 3 4 5 6 7     O1   4 3 1 2 5 6 7
         X     Y
P2   4 3 1 2 5 7 6     O2   2 1 3 4 5 7 6
```

Here the crossover points X and Y define an interchange mapping

$$3 \leftrightarrow 1$$

$$4 \leftrightarrow 2$$

$$5 \leftrightarrow 5$$

The results of PMX when applied to the TSP were satisfactory rather than spectacular, for reasons which will be discussed later.

Goldberg [2] describes two other types of crossover for sequence representations, while Reeves [27] used yet another to solve a flowshop sequencing problem. His C1 operator works by choosing a crossover point X randomly, taking the pre-X section of the first parent, and filling up the chromosome by taking in order each 'legitimate' element from the second parent. For the example above it might generate the following offspring:

```
P1   2 1 3 4 5 6 7        01   2 1 4 3 5 7 6
         X
P2   4 3 1 2 5 7 6        02   4 3 2 1 5 6 7
```

The rationale for C1 is that it preserves the absolute positions of the jobs taken from P1, and the relative positions of those from P2. It was conjectured that this would provide enough scope for modification of the chromosome without excessively disrupting it.

It would appear that this operator too has been independently 'discovered' more than once—Smith [51] would appear to be the first, but it has also been used by Prosser [52], and in an extended version by Oliver *et al.* [53], while more recently Davis [29] has described a generalization under the name of 'uniform order-based crossover'. The latter is based on the same idea as Syswerda's uniform crossover: a crossover template of 0s and 1s is randomly generated, where the 1s define elements taken from the first parent, while the other elements are copied from the second parent in the order in which they appear in that chromosome.

Thus applying the template

```
1 0 1 0 0 0 1
```

in the case above would have the following effect:

```
P1   2 1 3 4 5 6 7        01   2 4 3 1 5 6 7

P2   4 3 1 2 5 7 6        02   4 2 1 3 5 7 6
```

All these operators emphasize in one way or other the preservation of either relative or absolute positions in a string. Another approach has been taken by Whitley *et al.* [54] with their *edge-recombination* crossover which is described below in section 5.1. The thrust of this operator is rather to preserve *adjacency* of the components of the string.

Mutation also needs to be re-defined in the context of a sequence representation. Conventional mutation operates at each locus of the string in turn, which clearly has no parallels in the case of a permutation. Several possibilities exist: Reeves [27] explored an *exchange* mutation—the interchange of two randomly chosen elements of the permutation, and found it inferior to a *shift* mutation for flowshop sequencing. The latter variant was just the movement of a randomly chosen element a random number of places to the left or right.

Davis [29] also investigated the exchange type of mutation, but preferred a *scramble sublist* mutation. This acts by choosing two points on the string at random, and randomly permuting (scrambling) the elements between these two positions. Intuitively, this appears more disruptive than the others mentioned, and Davis found it necessary in some cases to limit the length of the portion which was scrambled.

4.4.4 Hybridization

In many cases, the effectiveness of a GA can be enhanced by *hybridizing* it with another heuristic. One of the main benefits of the GA paradigm is its *domain independence*; i.e. the fact that it operates on a coding of a problem, and needs to know nothing more about the problem itself. Creating a hybrid almost inevitably means that we lose this characteristic because some problem-specific knowledge is introduced, but this is often a price worth paying if our concern is to find an effective and efficient solution to a particular problem rather than to study GA performance in itself.

Hybridization can be carried out in various ways. The idea of seeding a population (as discussed above) with the results of applying a problem-specific heuristic can be viewed as a minimal-level hybridization. Several authors have described ways in which local neighbourhood search or extensions such as simulated annealing can be imbedded in a GA in order to enhance performance. One approach is to use a GA to find regions of the solution space which are good candidates for locating the optimum. In such cases, it is sensible to use speciation or other techniques to maintain a fair degree of population diversity.

Goldberg [2] describes an alternative method for incorporating neighbourhood search into a GA, in a procedure he calls *G-bit improvement.* With this approach, some of the best strings are periodically selected for a search of a neighbourhood defined by single bit reversals. On completion of the neighbourhood search, the locally optimal strings are re-introduced into the population for the next phase of the Genetic Algorithm. A similar version of this procedure has been proposed in [55], in which heuristic operators based on 2-optimal search and simulated annealing were used alongside simple crossover.

Another possibility [28, 56] is to use a GA to take the 'top-level' decisions on the form of a solution, which is then taken and solved by a problem-specific procedure. Again, it may be possible to devise special types of genetic operator relating to the problem in question, which are more powerful than simple crossover and mutation. Some of these are described below in the context of the applications in which they were introduced.

4.4.5 Parallel implementations

Quite apart from the intrinsic parallelism that Holland describes, there is clearly scope for implementing GAs in a parallel way. The obvious approach (assuming *en bloc* replacement) is to evaluate the fitness of each chromosome of the population in parallel. Provided there are enough processors, it should be possible to evaluate each generation in not much more time than a sequential program would take to evaluate an individual. However, this does create communication overheads which may limit the potential speed-up. Perhaps for this reason, most reported parallel implementations have taken another route, although the parallelization of fitness evaluations has been used by several authors [57, 58, 59]. In the paper by Talbi and Bessière [57], the communication overheads were limited by allocating individuals to processors, where mating was restricted to neighbouring individuals only. It is not entirely clear from their description, but it would appear that this approach was also taken by Mühlenbein *et al.* [58].

The second approach has been advocated by several authors [60,

61], under various guises. The essential feature is to allocate *sub-populations* of chromosomes to parallel processors which proceed independently for a certain number of generations. At this point, information is re-distributed among the sub-populations in some way—for instance, the best chromosome found so far amongst all sub-populations could be inserted into each sub-population. In fact Pettey *et al.* [60] took this route, by donating and receiving the best individuals once every generation, while Cohoon *et al.* [61] chose to copy randomly chosen subsets of solutions between sub-populations following a relatively large number of generations.

4.4.6 Computer software

Before proceeding to describe some of the applications, it is worth a brief digression into some practical matters relating to GA software. The basic concepts of a GA are actually fairly easy to program, but access to commercially available and public-domain software is also important. The book by Goldberg [2] contains a suit of Pascal programs which form a good starting point for a deeper understanding of GAs. Davis [29] also refers extensively to his own OOGA software, which is available for a modest charge, together with the GENESIS code developed by Grefenstette. The latter software is also available in more than one form from *ftp* archives in more than one place: for details on the latter, and much other useful material, prospective GA researchers are advised to join the electronic mailing list for GAs by contacting `GA-List-Request@aic.nrl.navy.mil`.

4.5 Applications

Most of the original applications of GAs were to areas of particular interest in the realm of Artificial Intelligence, and some of these have been referred to already. Recently, problems of interest in OR have also been tackled, and some of these are described below.

4.5.1 Travelling salesman problem

For genetic algorithms, as for any new approach for combinatorial optimization, the first problem that attracted attention was the travelling salesman problem (TSP).

As mentioned earlier, the natural representation of this problem is as a permutation, which implies the need for a modified crossover operator. The PMX operator of Goldberg and Lingle [50] was the first attempt to apply GAs to the TSP, in which they found near-optimal solutions to a well-known 33-city problem. However, this method took no account of the actual inter-city distance matrix, apart from the objective-function calculation.

In a later paper, Grefenstette [62] described an approach in which the GA is helped by incorporating problem-specific knowledge. The crossover operator used here was defined in the form of an iterative algorithm. Having selected two parents in the usual way, each city has up to 4 edges in the parent tours. One edge is chosen for the offspring at each iteration by first choosing a *city* randomly from the list of those not already used, and then choosing an edge incident at that city in accordance with a probability distribution that favours short edges. This procedure was supplemented by a mutation operator which used a form of rapid neighbourhood search, and the combined effect was to produce highly efficient tours. Nevertheless, the final tour could still often be improved by a neighbourhood search, and Suh and Van Gucht [55] have reported further improvements by incorporating heuristic operators based on 2-optimal and/or simulated annealing procedures.

Whitley *et al.* [54] have used a different method of defining crossover for the TSP. Grefenstette's edge-list concept, as originally introduced, led to frequent re-starts as the next city chosen had no more edges left in either parent. Whitley argued that this is undesirable, and that the choice of next city should instead be made on the basis of how many parent edges were available[2]. With this *edge-recombination* operator, several moderate to large problems were solved to optimality or to new best solutions.

4.5.2 Sequencing and scheduling

Another family of well-researched OR problems is that concerned with scheduling and sequencing of jobs and machines. Some of these are fairly easy to solve, but many are NP-hard. One of the most

[2]Readers may observe a connection here with Radcliffe's [23] idea of respect: Whitley's operator ensures that the offspring inherits edges which appear in both parents.

straightforward non-trivial problems in this family is the permutation flowshop-sequencing problem $n/m/P/C_{max}$. (For a guide to the standard terminology of sequencing problems, see for instance Rinnooy Kan [63].)

Reeves [27] formulated a GA for this problem which used the C1 crossover, an adaptive mutation rate, and a seeded population, building on a version of the simple GA used in another context by Ackley [31]. He compared its performance with that of a highly efficient simulated annealing heuristic [64]; over a large number of different-sized problems, the methods were broadly comparable, although it appeared that the GA did start to out-perform as the problem size increased. In another smaller-scale study [65], this GA was also compared to a simple tabu search, as well as to some naive neighbourhood search procedures. The GA easily out-performed the naive approaches, and produced similar results to tabu search.

In a further paper [66] the GA was applied to a *stochastic* flowshop problem. The assumption that job processing times are fixed deterministic values is highly unlikely in the real world, where a degree of uncertainty in the time it will take to process a job is often present. In such cases, sequencing on the basis of expected values would usually be adopted. Alternatively, one could try injecting some simulated 'noise' into the expected processsing times and use a heuristic designed for the deterministic case. Because of the nature of a GA, in which (good) schemata receive many trials during the course of a few generations, it is possible to inject some simulated 'noise' into the evaluation of each sequence without the fear that an 'outlier' will have undue influence on the sequence chosen. In [66] it was shown that introducing noise into other heuristics led to poor performance, whereas using a GA in the presence of noise actually led to substantial improvements over methods which relied on expected values.

Whitley *et al.* [54] have reported an extension of their edge-recombination operator to more general problems. While sequencing refers to the relative order or priority of jobs, *scheduling* relates to the problem of allocating these jobs to (parallel) machines, paying attention to any precedence relationships that exist. Scheduling is usually such a difficult problem that it has to be done by a problem-specific heuristic. In [54] this heuristic is 'fed' with a sequence so as to return a fitness value determined by the heuristic to the GA. This approach produced results comparable with hand-generated schedules, and fairly close to theoretical minima.

Beaty [56] has applied GAs to instruction scheduling for computers. This problem has many similarities to the job-shop scheduling problem, in that a computer program in a high-level language has to be translated into a set of instructions at the machine level; these instructions contend for resources and involve precedences. The basic scheme for instruction scheduling by GA was thus similar to that used in [54] for job-shop scheduling. Beaty compared different types of crossover operator, and concluded that while the edge-recombination operator performed well for pure *sequencing* problems, operators which preserve relative positions should be preferred for *scheduling*.

4.5.3 Graph colouring

Davis [29] has applied a genetic algorithm to the problem of colouring nodes on a graph—a problem which has been described earlier in Chapter 2. The approach taken was to hybridize a *greedy* algorithm for colouring the nodes: a permutation describes the order in which the nodes are examined. Each node is allocated the first colour that it can feasibly have, or none if its adjacent nodes do not permit it. The results of this procedure were greatly superior to a greedy algorithm in some experiments, but no comparisons with more effective methods such as simulated annealing were reported.

4.5.4 Steiner trees

Kapsalis *et al.* [28] have applied the GA approach to the problem of finding minimal Steiner trees on a graph. This is the problem of choosing a connected subgraph such that certain 'special vertices' are always included, in such a way as to minimize the sum of the edge-lengths of the sub-graph. In their heuristic, the GA was used to select a set of vertices for inclusion, following which a minimal spanning tree was found using a well-known standard algorithm. On a suite of benchmark problems, the optimal solution was found in every case; they also observed that the GA approach was very robust—many different values of the basic parameters were tried, with little variation in the quality of solution obtained.

4.5.5 Knapsack problems

Goldberg and Smith [45] described a GA for solving a *knapsack* problem. For the normal situation (as defined in chapter 1), the choice of chromosome representation is straightforward—a string of 0s and 1s, so that ordinary crossover could be applied. However, the problem considered by Goldberg and Smith (as discussed earlier) was non-stationary, which necessitated the use of a diploid representation. Apart from this, the interesting feature is the treatment of constraints. The obvious strategy is to create an augmented objective function by adding a penalty to the original one. Goldberg [2] recommends that the penalty should be found by squaring the amount by which a constraint is violated, but it is not clear that this is necessarily the best alternative.

Fairley [40] has carried out some extensive experiments on this aspect of knapsack problems, using his 'string-of-change' operator. The constraints were handled by multiplying the excess of constraint i by a factor λ_i. He found that when there are several constraints the choice of these penalty factors is critical in the performance of the GA. If they are set too high, the penalties dominate, and a zero solution results; too low and the procedure fails to eliminate the infeasibilities completely. A penalty factor based on links with Lagrangean Relaxation [67] was found to perform quite well but not spectacularly so.

4.5.6 Set covering problems

In [68, 69] Liepins and others have considered a number of GAs for set covering problems. As in the case of knapsack problems, the question of how to deal with constraints is a significant issue. In [68] it is pointed out that a poor choice of a penalty factor can have the effect of creating a 'deceptive' problem. Better choices of penalty factor require some examination of an infeasible solution in order to try to estimate what is needed to convert it to feasibility. Computing these factors is inevitably a problem-specific question, and the work reported confirms that solving constrained problems is not a simple matter. Nevertheless, Liepins and Potter [69] contend that with a good penalty function, a GA performs comparably with specialized OR techniques for set covering problems.

4.5.7 Bin packing

Bin packing provides another area where an 'obvious' GA implementation fails to perform. The problem involves the allocation of objects of different sizes to a set of bins in order to minimize the number of bins used. It is possible to formulate the problem in integer programming terms, and use a 0-1 coding, but this involves constraint violation problems similar to those arising in knapsack and set covering problems. Another coding would be to assign each object the bin number in which it is placed, but as Falkenauer and Delchambre [70] show, this also performs poorly. They develop an alternative combination of coding, fitness definition and crossover operator which produced very good results.

A different approach was taken by Prosser [52] in a similar problem where the requirement was to stack metal plates on pallets in order to minimize the number of pallets used. His GA used a sequence representation, in which the sequence represented the order in which the pallets would be loaded, on the understanding that a new pallet would be used whenever a loading constraint was violated. He also used modified crossover and inversion operators, applied in an iterative fashion. The first step was to load the first pallet with as much weight as possible; the GA was then used (having removed the stacks allocated to the first pallet) to load the second one, and so on. This method also produced excellent results.

4.5.8 Neural networks

Reeves and Steele [49] used a GA approach to the problem of designing the architecture and setting the parameters of an artificial neural network. This problem is often dealt with by trial-and-error, although it could be investigated more systematically by using a suitable experimental design, fitting a response surface, and attempting to optimize over this surface. However, the surface was expected to be discontinuous and non-linear—the type of situation in which such methods often perform badly, but where a GA should perform well. Although some of the parameters in this problem are continuous, it was decided to discretize them so that the whole problem became one of finding the best combination of values of a set of discrete variables—in other words a combinatorial problem. The method was tested on several problems, in which it proved decidedly superior to a 'try a random configuration and pick the best' approach. A similar method has been reported by Harp and Samad [71] who used a GA to analyse different network architectures in a very general way. Another example can be found in a paper by Bornholdt and Graudenz [72]. Their GA used mutation only, with no crossover, in a procedure which also allowed a very general architecture specification.

Montana [59] has also used a GA for training a network instead of the common *back-propagation* algorithm. The standard *multi-layer perceptron* architecture contains a set of parameters (called *weights*). When an input pattern is applied to the network a complicated non-linear transformation is computed in order to predict the corresponding (known) output value. Back-propagation adjusts the weights in order to reduce a function of the errors between actual and predicted outputs by performing a quasi-gradient descent in weight space, but it is equally possible to use a GA to find an appropriate set of weights. Montana developed several heuristic operators and tested them using numerical-valued coding of the chromosome; the best version of his GA outperformed backpropagation (for his problem) by a comfortable margin.

Other work on similar lines has also been reported by Marshall and Harrison [73], and in some rather more extensive experiments by Reeves and Steele [74]. The latter used only the standard binary coding with simple crossover and mutation; the results in terms of generalization performance (i.e. how a network trained on one set of data performs on an unseen test set) were much less erratic than networks trained by back-propagation. It is conjectured that this occurs because a GA is far less likely than back-propagation to find a poor local optimum in weight space.

4.5.9 Other problems

Goldberg [2] describes two applications to optimization in engineering. One, a problem of finding a flow schedule for a natural gas pipeline, was essentially a complicated non-linear programming problem, and the results compared very favourably with more traditional methods. A quadratic penalty function was used here to deal with constraints.

Another non-linear optimization problem which he approached

via a GA was that of minimizing the weight of a truss subject to stress constraints. Other methods were available for the particular problem reported, so that the GA could be bench-marked. Later research suggested that the GA was capable of solving problems which conventional methods find difficult.

Glover [75] reported a successful application of a GA to a problem of keyboard configuration for East Asian languages. Unlike Western languages, the 'alphabet' is not a small set of letters, but a much larger set of basic components or *primitives* which when combined in an appropriate sequence define a character or *ideogram*—one of the basic 'words' of the language. The three most commonly used GA operators—crossover, inversion and mutation—were modified somewhat in this context, and the resulting algorithm seemed to produce very satisfactory results. However, Glover did not compare this method with any other strategies.

Finally, it is worth mentioning the use of GAs as an alternative to conventional methods of experimental design. Grefenstette [76] applied genetic search methods to the task of designing a GA. Most GAs need some form of control mechanism, in order to define crossover and mutation rates, population sizes, selection mechanisms and so on. It is far from clear what settings these parameters should have, although experimental studies have provided some guidance. Grefenstette designed a *meta-GA* which was able to optimize GA performance in parameter space over several different functions.

4.6 Conclusions

This chapter has attempted to introduce the reader to the fast-developing field of genetic algorithms, and it should be realised that although it has concentrated mainly on the question of how GAs relate to combinatorial optimization, there is a large volume of research in GAs that relates to other areas. Readers who want to find out more are recommended to consult the books by Goldberg [2] and Davis [29] in the first place. However, GA research is undergoing exponential growth, and to keep up with events, it will be necessary to consult the proceedings of various conferences and workshops. Again, the best place to find out about these events is the regular GA Digest available by electronic mail from the address quoted earlier.

In summary, the principal attractions of GAs are:

- *Domain independence* The algorithms work on a *coding* of a problem, so that it is easy to write one general computer program for solving many different optimization problems.

- *Non-linearity* Many conventional optimization techniques rely on unrealistic assumptions of linearity, convexity, differentiability etc. None of these are needed by a genetic algorithm; the only requirement is the ability to calculate some measure of performance, which may be highly complicated and non-linear.

- *Robustness* As a consequence of these two characteristics, GAs are inherently robust—they can cope with a diversity of problem types, they can work with highly non-linear functions, and they do it, as has been shown earlier, in a very efficient manner. Further, empirical evidence is strong that although it is possible to fine-tune a GA to work better on a given problem, it is nonetheless true that a wide range of parameter settings (population sizes, crossover and mutation rates etc.) will give very acceptable results.

- *Ease of modification* Even relatively minor modifications to a particular problem may cause severe difficulties to many heuristics. By contrast, it is easy to change a GA to model variations of the original problem.

- *Parallel nature* Quite apart from the property of intrinsic parallelism which GAs have been shown to possess, there is great scope for *implementing* GAs in parallel. As more sophisticated computing hardware becomes available, developments in this area will surely become increasingly common.

Despite these important features, the GA paradigm is not a panacea, and indeed, may often be out-performed by purpose-built methods in a particular context. There is sometimes a tendency among GA enthusiasts to see applications everywhere, precisely because of the advantages discussed above, but it usually makes sense to incorporate problem-specific information, perhaps by hybridizing. Another related criticism that may justly be levelled at much of the research in GAs is that there has sometimes been insufficient benchmarking of performance *vis-à-vis* other heuristic methods, although what has been reported suggests that they would not suffer from such a comparison.

There are clearly several outstanding problems to be resolved in the application of GAs (quite apart from the problems of their theoretical analysis, which is lagging some way behind their practice). One of the most important questions to be addressed is how to ensure that the coding of the problem is such that the building-block hypothesis is valid—or, alternatively, how to adapt the standard crossover operator if it is not. (Further developments in the implementation of messy GAs may be a solution.) As pointed out in chapter 3, a difficulty here is that the GA's independence of its context (an advantage in some respects) can sometimes handicap the algorithm severely, and the creation of hybrids with tabu search may be important in overcoming such problems. Genetic algorithms are highly efficient methods for *intensification*, to use the language of tabu search, but the incorporation of *diversification* strategies may increase their capacity for solving difficult problems. Indeed, there are many other aspects of tabu search which find an echo in GAs, and the development and exploitation of some of the frontier areas outlined in chapter 3 may have a significant impact in both areas.

Nevertheless, across a wide spectrum of problems, it is usually true that even a simple genetic algorithm will probably perform as well as, or better than, most other general techniques, and it is hoped that sufficient has been described here to make OR workers aware of a significant new weapon in their armoury for solving optimization and search problems.

References

[1] J.H.Holland (1975) *Adaptation in Natural and Artificial Systems.* University of Michigan Press, Ann Arbor.

[2] D.E.Goldberg (1989) *Genetic Algorithms in Search, Optimization, and Machine Learning.* Addison-Wesley, Reading, Mass.

[3] S.M.Roberts and B.Flores (1966) An engineering approach to the travelling salesman problem. *Man.Sci.*, **13**, 269-288.

[4] C.E.Nugent, T.E.Vollman and J.E.Ruml (1968) An experimental comparison of techniques for the assignment of facilities to locations. *Ops.Res.*, **16**, 150-173.

[5] G.E.P.Box and N.R.Draper (1987) *Empirical Model-Building and Response Surfaces.* Wiley, Chichester.

[6] J.D.Bagley (1967) *The behavior of adaptive systems which employ genetic and correlation algorithms.* Doctoral dissertation, University of Michigan.

[7] R.S.Rosenberg (1967) *Simulation of genetic populations with biochemical properties.* Doctoral dissertation, University of Michigan.

[8] D.J.Cavicchio (1972) *Adaptive search using simulated evolution.* Doctoral dissertation, University of Michigan.

[9] J.Antonisse (1989) A new interpretation of schema notation that overturns the binary encoding constraint. *In* [10], 86-91.

[10] J.D.Schaffer (Ed.) (1989) *Proceedings of 3rd International Conference on Genetic Algorithms.* Morgan Kaufmann, Los Altos, CA.

[11] J.D.Schaffer (1987) Some effects of selection procedures on hyperplane sampling by genetic algorithms. *In* [12], 89-103.

[12] L.Davis (Ed.) (1987) *Genetic Algorithms and Simulated Annealing.* Morgan Kauffmann, Los Altos, CA.

[13] C.L.Bridges and D.E.Goldberg (1987) An analysis of reproduction and crossover in a binary-coded genetic algorithm. *In* [14], 9-13.

[14] J.J.Grefenstette(Ed.) (1987) *Proceedings of the 2nd International Conference on Genetic Algorithms.* Lawrence Erlbaum Associates, Hillsdale, NJ.

[15] K.G.Beauchamp (1975) *Walsh Functions and Their Applications.* Academic Press, London.

[16] A.D.Bethke (1981) *Genetic algorithms as function optimizers.* Doctoral dissertation, University of Michigan.

[17] D.E.Goldberg (1989) Genetic algorithms and Walsh functions: part I, a gentle introduction. *Complex Systems*, **3**, 129-152.

[18] D.E.Goldberg (1989) Genetic algorithms and Walsh functions: part II: deception and its analysis. *Complex Systems*, **3**, 153-171.

[19] D.E.Goldberg and M.Rudnick (1991) Genetic algorithms and the variance of fitness. *Complex Systems*, **5**, 265-278.

[20] D.Whitley(1992) Deception, dominance and implicit parallelism in genetic search. *Annals of Maths. and AI*, **5**, 49-78.

[21] D.E.Goldberg, B.Korb and K.Deb (1989) Messy genetic algorithms: motivation, analysis and first results. *Complex Systems*, **3**, 493-530.

[22] D.E.Goldberg, K.Deb and B.Korb (1990) Messy genetic algorithms revisited: studies in mixed size and scale. *Complex Systems*, **4**, 415-444.

[23] N.J.Radcliffe (1991) Equivalence class analysis of genetic algorithms. *Complex Systems*, **5**, 183-205.

[24] K.A.De Jong (1975) *An analysis of the behavior of a class of genetic adaptive systems.* Doctoral dissertation, University of Michigan.

[25] D.E.Goldberg (1989) Sizing populations for serial and parallel genetic algorithms. *In* [10], 70-79.

[26] J.T.Alander (1992) On optimal population size of genetic algorithms. *Proc. CompEuro 92*, 65-70. IEEE Computer Society Press.

[27] C.R.Reeves (1992) A genetic algorithm for flowshop sequencing. *Computers & Ops.Res.*, (in review).

[28] A.Kapsalis, G.D.Smith and V.J.Rayward-Smith (1993) Solving the graphical steiner tree problem using genetic algorithms. *JORS*, (to appear).

[29] L.Davis (Ed.) (1991) *Handbook of Genetic Algorithms.* Van Nostrand Reinhold, New York.

[30] D.Whitley (1989) The GENITOR algorithm and selection pressure: why rank-based allocation of reproductive trials is best. *In* [10], 116-121.

[31] D.H.Ackley (1987) An empirical study of bit vector function optimization. *In* [12], 170-204.

[32] L.B.Booker (1982) *Intelligent behavior as an adaptation to the task environment.* Doctoral dissertation, University of Michigan.

[33] A.Brindle (1981) *Genetic algorithms for function optimization.* Doctoral dissertation, University of Alberta, Edmonton.

[34] J.E.Baker (1987) Reducing bias and inefficiency in the selection algorithm. *In* [14], 14-21.

[35] D.E.Goldberg and J.Richardson (1987) Genetic algorithms with sharing for multimodal function optimization. *In* [14], 41-49.

[36] R.B.Hollstien (1971) *Artificial genetic adaptation in computer control systems.* Doctoral dissertation, University of Michigan.

[37] J.E.Baker (1985) Adaptive selection methods for genetic algorithms. *In* [38], 101-111.

[38] J.J.Grefenstette(Ed.) (1985) *Proc. of an International Conference on Genetic Algorithms and their applications.* Lawrence Erlbaum Associates, Hillsdale, NJ.

[39] L.B.Booker (1987) Improving search in genetic algorithms. *In* [12], 61-73.

[40] A.Fairley (1991) *Comparison of methods of choosing the crossover point in the genetic crossover operation.* Dept. of Computer Science, University of Liverpool.

[41] L.J.Eshelman, R.A.Caruana and J.D.Schaffer (1989) Biases in the crossover landscape. *In* [10], 10-19.

[42] G.Syswerda (1989) Uniform crossover in genetic algorithms. *In* [10], 2-9.

[43] D.J.Sirag and P.T.Weisser (1987) Towards a unified thermodynamic genetic operator. *In* [14], 116-122.

[44] K.A.De Jong and W.M.Spears (1992) A formal analysis of the role of multi-point crossover in genetic algorithms. *Annals of Maths. and AI*, **5**, 1-26.

[45] D.E.Goldberg and R.E.Smith (1987) Nonstationary function optimization using genetic algorithms with dominance and diploidy. *In* [14], 59-68.

[46] I.Harvey (1992) *Evolutionary robotics and SAGA: the case for hill crawling and tournament selection.* School of Cognitive and Computing Sciences, University of Sussex.

[47] J.D.Schaffer, R.A.Caruana, L.J.Eshelman and R.Das (1989) A study of control parameters affecting online performance of genetic algorithms for function optimization. *In* [10].

[48] R.A.Caruana and J.D.Schaffer (1988) Representation and hidden bias: Gray vs. binary coding for genetic algorithms. *Proc. 5th International Conference on Machine Learning.* Morgan Kaufmann, Los Altos, CA.

[49] C.R.Reeves and N.C.Steele (1991) A genetic algorithm approach to designing neural network architecture. *Proc. 8th International Conference on Systems Engineering.*

[50] D.E.Goldberg and R.Lingle (1985) Alleles, loci and the travelling salesman problem. *In* [38], 154-159.

[51] D.Smith (1985) Bin packing with adaptive search. *In* [38], 202-206.

[52] P.Prosser (1988) A hybrid genetic algorithm for pallet loading. *Proc. 8th European Conference on Artificial Intelligence.* Pitman, London.

[53] I.M.Oliver, D.J.Smith and J.R.C.Holland (1987) A study of permutation crossover operators on the traveling salesman problem. *In* [14], 224-230.

[54] D.Whitley, T.Starkweather and D.Shaner (1991) The traveling salesman and sequence scheduling: quality solutions using genetic edge recombination. *In* [29], 350-372.

[55] J.Y.Suh and D.Van Gucht (1987) Incorporating heuristic information into genetic search. *In* [14], 100-107.

[56] S.Beaty (1991) *Instruction scheduling using genetic algorithms.* Doctoral dissertation, Colorado State University.

[57] E.G.Talbi and P.Bessière (1991) A parallel genetic algorithm applied to the mapping problem. *SIAM News*, July 1991, 12-27.

[58] H.Mühlenbein, M.Gorges-Schleuter and O.Krämer (1988) Evolution algorithms in combinatorial optimization. *Parallel Computing*, **7**, 65-85.

[59] D.J.Montana (1991) Automated parameter tuning for interpretation of synthetic images. *In* [29], 282-311.

[60] C.B.Pettey, M.R.Leuze and J.J.Grefenstette (1987) A parallel genetic algorithm. *In* [14], 155-161.

[61] J.P.Cohoon, S.U.Hegde, W.N.Martin and D.Richards (1987) Punctuated equilibria: a parallel genetic algorithm. *In* [14], 148-154.

[62] J.J.Grefenstette (1987) Incorporating problem-specific knowledge into genetic algorithms. *In* [12], 42-60.

[63] A.H.G.Rinnooy Kan (1976) *Machine Scheduling Problems: Classification, Complexity and Computations.* Martinus Nijhoff, The Hague.

[64] I.H.Osman and C.N.Potts (1989) Simulated annealing for permutation flowshop scheduling. *OMEGA*, **17**, 551-557.

[65] C.R.Reeves (1991) Recent algorithmic developments applied to scheduling problems. *Proc. 9th IASTED Symposium on Applied Informatics.* ACTA Press, Anaheim, CA.

[66] C.R.Reeves (1992) A genetic algorithm approach to stochastic flowshop sequencing. *Proc. IEE Colloquium on Genetic*

Algorithms for Control and Systems Engineering. Digest No.1992/106, IEE, London.

[67] J.G.Klinciewicz and H.Luss (1986) A Lagrangian relaxation heuristic for capacitated facility location with single-source constraints. *JORS*, **37**, 495-500.

[68] J.T.Richardson, M.R.Palmer, G.E.Liepins and M.R.Hillyard (1989) Some guidelines for genetic algorithms with penalty functions. *In* [10], 191-197.

[69] G.E.Liepins and W.D.Potter (1991) A genetic algorithm approach to multiple-fault diagnosis. *In* [29], 237-250.

[70] E.Falkenauer and A.Delchambre (1992) A genetic algorithm for bin packing and line balancing. *Proc.IEEE International Conference on Robotics and Automation.*

[71] S.A.Harp and T.Samad (1991) Genetic synthesis of neural network architecture. *In* [29], 202-221.

[72] S.Bornholdt and D.Graudenz (1992) General asymmetric neural networks and structure design by genetic algorithms. *Neural Networks*, **5**, 327-334.

[73] S.J.Marshall and R.F.Harrison (1991) Optimization and training of feedforward neural networks by genetic algorithms. *Proc. 2nd IEE International Conference on Artificial Neural Networks*, 39-43.

[74] C.R.Reeves and N.C.Steele (1992) Problem-solving by simulated genetic processes: a review and application to neural networks. *Proc. 10th IASTED Symposium on Applied Informatics.* ACTA Press, Anaheim, CA.

[75] D.E.Glover (1987) Solving a complex keyboard configuration problem through generalized adaptive search. *In* [12], 12-31.

[76] J.J.Grefenstette (1986) Optimization of control parameters for genetic algorithms. *IEEE-SMC*, **SMC-16**, 122-128.

Chapter 5

Artificial Neural Networks

Carsten Peterson and Bo Söderberg

5.1 Introduction

There has been an upsurge in interest in artificial neural networks (ANN) over the last 5 years, mostly in the area of feed-forward architectures for pattern recognition applications. There are several reasons for this. One is that some theoretical obstacles with the so-called perceptron, which was popular in the late 1960s, have been overcome. Also the recently available inexpensive access to powerful CPUs has facilitated the development of the field both with respect to model-building, and, very importantly, to dealing with real-world problems. Indeed, results from impressive 'product' quality application work keeps appearing both from the commercial and academic sectors.

Two main kinds of architecture exist, *feed-forward* and *feed-back*, with major application areas of feature recognition and optimization, respectively. The feed-forward ANN approach to classification problems constitutes, in a sense, extensions of 'conventional' fitting and principal component analysis tools into the non-linear domain. This is in contrast to using feed-back ANN to find solutions to difficult combinatorial optimization problems. The neural approach to these problems really brings something new to the table. It is not just a question of mapping the problem onto an energy function of discrete neural degrees of freedom. Using mean field (MFT) equations to find the lowest energy is a novel technique for Operational Research borrowed from physics. Furthermore these MFT equations are isomorphic to VLSI RC-equations, making the hardware implementation straightforward.

This chapter is intended as a self-contained introduction to using ANN for finding good approximate solutions to difficult optimization problems. It is organized as follows. Section 5.2 gives a general introduction to neural systems in general, biological neural networks and the abstraction to artificial neural networks. In section 5.3 an overview is presented of the problem of using ANN for combinatorial optimization problems. The graph bisection problem, which is particularly suited for binary neurons is dealt with in section 5.4. Multi-state or Potts neurons are introduced and illustrated with the graph partition problem in section 5.5 and the travelling salesman problem in section 5.6. The Potts formulation is very suitable for scheduling problems, which are discussed in section 5.7. An alternative procedure for geometrical low dimensional problems, the deformable templates approach, is described in section 5.8 in the context of the travelling salesman problem. Using ANN to find solutions to resource allocation (knapsack) problems is described in section 5.9. Section 5.10 contains a brief summary and outlook.

5.2 Neural Networks

An artificial neural network is a computational paradigm that differs substantially from those based on the 'standard' von Neumann architecture. Feature recognition ANNs generally learn from 'experience', instead of being expressly 'programmed' with rules as in conventional artificial intelligence (AI). Feedback ANNs 'feel' their way to good solutions, rather than explicitly exploring possibilities.

5.2.1 Biological neural networks

The ANN is inspired by the structure of biological neural networks and their way of encoding and solving problems. We will here very briefly review the basic components and functionality of the vertebrate central nervous system (CNS), whose details were revealed around 1940 with the emergence of the electron microscope. For more extensive literature on this subject we refer the reader to [1].

The human brain contains approximately 10^{12} *neurons*. These can be of many different types, but most of them have the same general structure (see Figure 5.1). The *cell body* or *soma* receives electric input signals to the *dendrites* by means of ions. The interior of the cell body is negatively charged against a surrounding *extra-cellular fluid.* Signals arriving at the dendrites depolarize the resting potential, enabling Na^+ ions to enter the cell through the membrane, resulting in an electric discharge from the neuron—the neuron 'fires'. The accumulated effect of several simultaneous signals arriving at the dendrites is usually approximately linearly additive whereas the resulting output is a strongly non-linear all-or-none type process. The discharge propagates along the *axon* to a *synaptic junction*, where *neurotransmitters* travel across a *synaptic cleft* and reach the dendrites of the postsynaptic neuron. A synapse which repeatedly triggers the activation of a postsynaptic neuron will grow in strength; others will gradually weaken. This *plasticity*, which is known as the *Hebb rule*, plays a key part in *learning.*

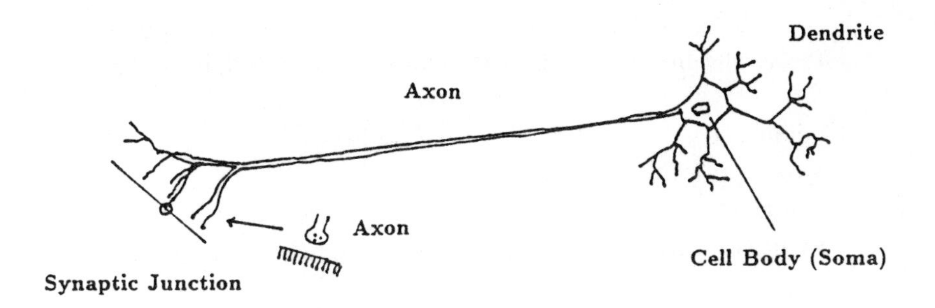

Figure 5.1: Schematic structure of a neuron

This general neuron structure is implemented in many different sizes and forms with different functionalities. Cell bodies have diameters in the range 5-80 μm and dendrite 'trees' extend from 10 μm up to 2-3 mm. Axons can be up to 1 m in length, while sizes of entire neurons vary from 0.01 mm to 1 m. *Primary sensory neurons* connect muscles or receptors to neurons, *secondary sensory neurons* and *interneurons* connect neurons with neurons, while *motor neurons* connect neurons with muscle fibres.

The *connectivity* (number of neurons connected to a neuron) varies from ~ 1 to $\sim 10^5$. For the *cerebral cortex* $\sim 10^3$ is an average. This corresponds to $\sim 10^{15}$ synapses per brain. Synapses can be either *excitatory* or *inhibitory* of varying strength. In the simplified binary case of just two states per synapse the brain thus has $\sim 2^{10^{15}} \approx 10^{10^{14}}$ possible configurations! The neural network is consequently in sharp contrast to a von Neumann computer with respect both to architecture and to functionality (see Table 5.1).

Table 5.1: Comparison of characteristics of neural networks and conventional computers

Vertebrate brain (Parallel distributed proc.)	Conventional computer (von Neumann machine)
Power dissipation: $\approx$ 100W	Power dissipation: $\approx 10^5$W
Good at: • recognition • adaptation • optimization (rough)	Good at: • $a + b$, $a \cdot b$, ... • if ... then ... else ... • state space exploration
$\tau = \mathcal{O}(\text{ms})$	$\tau = \mathcal{O}(\text{ns})$
Parallel Robust Fault Tolerant	Serial/parallel Fragile

The von Neumann computer was originally developed for 'heavy duty' numerical computing but has later also turned out to be profitable for data handling, word processing etc. However, when it comes to matching the vertebrate brain in terms of performing 'human' tasks it has very strong limitations. There are therefore strong reasons for designing an architecture and algorithm that resembles more closely that of the vertebrate brain.

5.2.2 Artificial neural networks

The philosophy of the ANN approach is to abstract some key ingredients from biology, and out of these construct simple mathematical models that exhibit most of the above-mentioned appealing features[1]. In physics one has good experience of model-building out of major abstractions. For example, details of individual atoms in a solid can be lumped into effective 'spin' degrees of freedom in such a way that good descriptions of *collective* phenomena (phase transitions etc.) are obtained. It is exactly the collective behavior of the neurons that is interesting from the point of view of intelligent data processing.

Basics

The basic computational entities of an ANN are the neurons v_i, which can take real values within the interval [0,1] (or [-1,1]). Sometimes the even simpler binary neuron s_i is used, where $s_i = \{0, 1\}$ (or $\{-1, 1\}$).

These are of course simplifications of the biological neurons described in the previous section. Common to most neural models is a *local updating rule* (see Figure 5.2a)

$$v_i = g(\sum_j \omega_{ij} v_j - \theta_i) \tag{5.1}$$

where v_j are all neurons that are feeding to neuron v_i, through weights (synapses) ω_{ij}. These weights can have both positive (excitatory) and negative (inhibitory) values. The θ_i term is a threshold, corresponding to the membrane potential in a biological neuron. The non-linear *transfer function* $g(\cdot)$ is typically a sigmoid-shaped function such as

$$g(x) = 0.5[1 + \tanh(x/T)] \tag{5.2}$$

where the 'temperature' T sets the inverse gain; a low temperature corresponds to a very steep sigmoid and a high temperature corresponds to an approximately constant $g(\cdot)$ (see Figure 5.2b). The limit $(T \to 0)$ corresponds to binary neurons s_i .

This simple artificial neuron mimics the main features of real biological neurons in terms of linear additivity for the inputs and strong non-linearity for the resulting output. If the integrated input signal is larger than a certain threshold θ_i the neuron will 'fire'.

[1] For a general textbook on ANNs see e.g. [2].

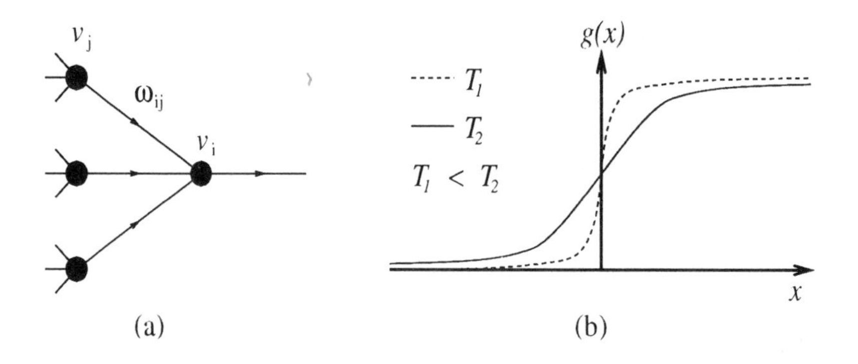

Figure 5.2: (a) Neuron updating and (b) sigmoid response functions of eqs. (5.1,5.2) for different temperatures T

There are two different kinds of architectures in neural network modelling; feed-forward (Figure 5.3a) and feed-back (Figure 5.3b). In feed-forward networks signals are processed from a set of input units in the bottom to output units in the top, layer by layer, using the local updating rule of equation (5.1). In feed-back networks, on the other hand, the synapses are bidirectional. Activation continues until a fixed point has been reached, reminiscent of a statistical mechanics system.

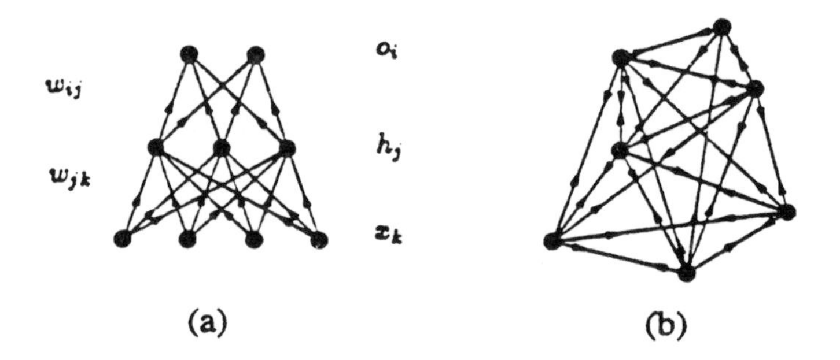

Figure 5.3: (a) Feed-forward and (b) feed-back architectures

Application areas

How can the power of these networks be exploited? There are four major application areas with soft borderlines in between—feature recognition, function approximation, associative memories and optimization. Feature recognition is the most exploited application domain to date.

Feature recognition In feature recognition (or pattern classification) situations one wants to categorize a set of input patterns $\vec{x}^{(p)} = (x_1, x_2, \ldots, x_N)^{(p)}$ in terms of different features o_i. The input patterns are fed into an input layer (receptors) and the output nodes represent the features. For the architecture depicted in Figure 5.3a, o_i depends on $\vec{x}^{(p)}$ through the functional form (cf. equation (5.1))

$$o_i(\vec{x}^{(p)}) = g(\sum_{j=0} \omega_{ij} g(\sum_{k=0} \omega_{jk} x_k^{(p)})) \tag{5.3}$$

where ω_{ij} and ω_{jk} are the parameters. Equation (5.3) can of course be generalized to any number of layers. (It should be noted that the thresholds, or bias terms, θ_i appearing in equation (5.1) have in equation (5.3) been transformed into weights ω_{i0} by adding an extra 'dummy' unit v_0 to each layer, which is constantly firing.) Fitting ω_{ij} and ω_{jk} to a given data set (or *learning*) takes place with gradient descent on a suitable *error function*. In this process the *training patterns* are presented over and over again with successive adjustments of the weights (*back-propagation* of the error [3]). Once this iterative learning has reached an acceptable level, in terms of a low error, the weights are frozen and the ANN is ready to be used on patterns it has never seen before. The capability of the network correctly to characterize these *test patterns* is called *generalization* performance. This procedure is equivalent to 'normal' curve fitting where a smooth parameterization from a training set is used to interpolate between the data points (generalization).

Function approximation Rather than having a 'logical' output unit (o_i) with a threshold behaviour described by equations (5.1, 5.2), one could imagine having an output representing a real number, corresponding to replacing equations (5.1, 5.2) with a linear behaviour. In this case one adjusts the weight parameters to parameterize an unknown real-valued function. Such an approach can be useful in *time-series predictions*, where one aims at predicting future values of a series given previous values [4, 5], e.g.

$$x_t = \mathcal{F}(x_{t-1}, x_{t-2}, ...) \tag{5.4}$$

where x_t is the real-valued output node and $x_{t-1}, x_{t-2}, ...$ are the values of the series at previous times.

Associative memory In this application there is no input/output distinction. Rather the network learns a set of identity mappings by fitting ω_{ij} to memory patterns or 'words' $\vec{s} = (s_1, s_2, s_3, ... s_N)$. If an incomplete or partly erroneous memory is presented to the network, it completes the pattern (see e.g. [6]). In a certain sense, feature recognizers are special cases of associative memories.

Optimization Feed-back networks have shown great promise in finding good solutions to difficult combinatorial optimization problems. In this case, again, non-linear updating equations (equation (5.1)) are allowed to settle. This corresponds to minimizing an energy function, which typically looks like

$$E = -\frac{1}{2} \sum_{i,j} \omega_{ij} s_i s_j \tag{5.5}$$

The problem is mapped onto the form of equation (5.5) by a clever choice of ω_{ij}. In this case ω_{ij} are fixed once and for all for each problem—they are not adaptive as in the other domains of application. The neurons s_i are then allowed to settle into a stable state, where the solution to the problem is given by the configuration $\vec{s} = (s_1, s_2, ...)$ with minimum energy.

5.3 Combinatorial Optimization Problems

Many combinatorial optimization problems are NP-complete. As discussed in chapter 1, exact solutions to these problems require state space exploration of one kind or another leading to $n!$ or a^n computations for a system with n degrees of freedom. Different kinds of heuristic methods are therefore often used to find reasonably good solutions. The ANN approach, which is based on feed-back networks, falls within this category. It has advantages in terms of solution quality and a 'fuzzy' interpretation of the answers through the MFT variables. Furthermore it is inherently parallel, facilitating implementations on concurrent processors, and with MFT equations custom-made hardware is straightforward to design. The ANN approach differs from exact and most other heuristic methods in the sense that it has no trial-and-error mechanism—it 'feels' its way to a good solution.

There are two families of ANN algorithms for optimization problems:

- *the 'pure' neural approach* based on either binary [7] or multistate [8] neurons with MFT equations for the dynamics;
- *deformable templates* [9], where the neural degrees of freedom have been integrated out and one is left with coordinates for possible solutions.

The latter pathway is a winner for low-dimensional geometrical problems like the travelling salesman problem (TSP), whereas the former is the only possibility for high-dimensional problems like scheduling. The following sections illustrate both of these approaches with suitable problems.

5.4 The Graph Bisection Problem

The neural approach is very transparent in this problem, since it is of a binary nature. We will here go systematically through the three basic steps which are common to all problems:

- map the problem onto a neural network;
- derive the MFT equations;
- analyze the dynamics of the MFT equations.

In addition we will of course test the approaches numerically.

5.4.1 Neural mapping

The graph bisection problem (GBP) is defined as follows (see Figure 5.4a): Partition a set of N nodes with given connectivity into two

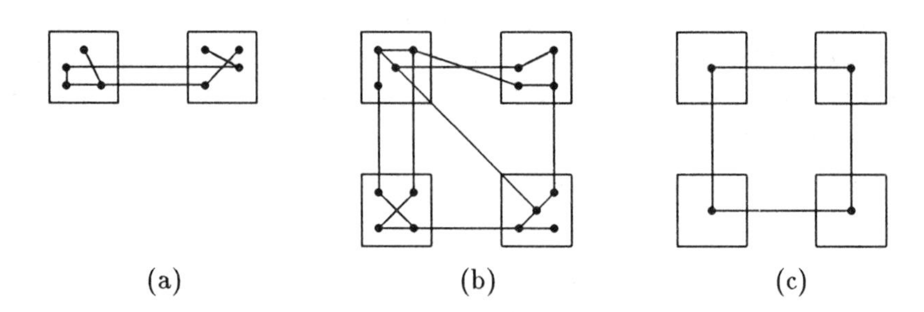

Figure 5.4: (a) A graph bisection problem (b) A $K = 4$ graph partition problem (c) A N=4 TSP problem

halves such that the connectivity (cutsize) between the two halves is minimized.

The problem is mapped onto the Hopfield energy function (cf. equation (5.5)) by the following representation: For each node, we assign a binary neuron $s_i = \pm 1$ and for each pair of vertices $s_i s_j$ $(i \neq j)$ we assign a value $\omega_{ij} = 1$ if they are connected, and $\omega_{ij} = 0$ if they are not. In terms of Figure 5.4a, we let $s_i = \pm 1$ represent whether node i is in the left or in the right position. With this notation,

$$\omega_{ij} s_i s_j \Rightarrow \begin{cases} > 0 & \text{whenever } i \text{ and } j \text{ are in the same partition} \\ = 0 & \text{if } i \text{ and } j \text{ are not connected at all} \\ < 0 & \text{whenever } i \text{ and } j \text{ are in different partitions} \end{cases} \tag{5.6}$$

Minimization of the energy function (5.5) will maximize the connections within a partition while minimizing the connections between partitions, with the result that all nodes are forced into one partition. Hence we must add a 'constraint term' to the right hand side of equation (5.5) that penalizes situations where the nodes are not equally partitioned. We note that $\sum s_i = 0$ when the partitions are balanced and a term proportional to $(\sum s_i)^2$ will thus increase the energy whenever the partition is unbalanced. Our energy function for graph bisection thus takes the form:

$$E = -\frac{1}{2}\sum_{ij}\omega_{ij}s_i s_j + \frac{\alpha}{2}(\sum_i s_i)^2 \tag{5.7}$$

where the *constraint coefficient* (imbalance parameter) α sets the relative strength between the cutsize and the balancing term. This balancing term represents a *global constraint.* The generic form of equation (5.7) is

$$E = \text{'cost'} + \text{'global constraint'} \tag{5.8}$$

which is typical when casting combinatorial optimization problems onto neural networks. The origin of the difficulty inherent in these kinds of problems is very transparent here: the problem is frustrated in the sense that the two constraints ('cost' and 'global constraint') are competing with each other. This leads to the appearance of many local minima. Similar situations occur in spin-glass systems in physics, which are very closely related to artificial neural networks.

The next step will be to find an efficient procedure for minimizing equation (5.7), such that local minima are avoided as much as possible. The MFT equations will turn out to be very powerful in this respect.

5.4.2 The mean field equations

A straightforward method for minimizing equation (5.7) would be to update s_i according to a local optimization rule,

$$s_i = sgn\left(\sum_j(\omega_{ij} - \alpha)\right) \tag{5.9}$$

With this procedure the system typically ends up in the local minimum closest to the starting point, which is not desired. Thus, a stochastic algorithm is needed that allows for uphill moves. One such procedure is simulated annealing (SA) [10], which has been extensively described in chapter 2. In this context, SA consists of generating, via neighbourhood search methods, configurations obeying the Boltzmann distribution

$$P[s] = \frac{1}{Z}e^{-E[s]/T} \tag{5.10}$$

where Z is the *partition function*

$$Z = \sum_{[s]} e^{-E[s]/T} \tag{5.11}$$

and T is the temperature of the system (for a general introduction to statistical mechanics, see e.g. Chandler [11]). By generating configurations at successively lower values of T (annealing) they are less likely to get stuck in a bad local minimum. Needless to say such a procedure can be very CPU consuming.

The MFT approach aims at approximating the stochastic SA method with a set of deterministic equations. The derivation has two steps. Firstly, the partition function of equation (5.11) is rewritten in terms of an integral over new continuous variables u_i and v_i. Secondly, Z is approximated by the maximum value of its integrand.

To this end, we embed the spins s_i in a linear space (in this case $\mathcal{R}$), introduce a new set of variables v_i existing in this space, one for each spin, and set them equal to the spins with a Dirac delta-function. Then we can express the energy in terms of the v_i, and Z takes the form

$$Z = \sum_{[s]} \int d[v] e^{-E[v]/T} \prod_i \delta(s_i - v_i) \tag{5.12}$$

Now, we Fourier-expand the delta-functions, introducing a set of conjugate variables u_i:

$$Z = \sum_{[s]} \int d[v] \int d[u] e^{-E[v]/T} \prod_i e^{u_i(s_i - v_i)} \tag{5.13}$$

then carry out the original sum over $[s]$:

$$Z \propto \int d[v] \int d[u] e^{-E[v]/T - \sum_i u_i v_i + \sum_i \log\cosh u_i} . \tag{5.14}$$

The original partition function is now rewritten entirely in terms of the new variables $[u, v]$, with an effective energy in the exponent. So far no approximation has been made. We next assume that Z in equation (5.13) can be approximated by the maximal value of the integrand—the saddlepoint approximation. For the position of the saddlepoint we obtain

$$u_i = -\frac{\partial E[v]}{\partial v_i}/T , \tag{5.15}$$

$$v_i = \tanh u_i . \tag{5.16}$$

Combining equations (5.15, 5.16) we obtain

$$v_i = \tanh\left(-\frac{\partial E[v]}{\partial v_i}/T\right) . \tag{5.17}$$

which for the energy of equation (5.7) reads

$$v_i = \tanh\left(\sum_j (\omega_{ij} - \alpha) v_j / T\right) . \tag{5.18}$$

The mean field variables v_i can be interpreted as $\langle s_i \rangle_T$, i.e. as thermal averages of the original binary spins. We thus recover the local updating equation (equations (5.1, 5.2)). What we have obtained is a set of deterministic equations emulating the stochastic behaviour. High temperatures correspond to very smooth sigmoids and the low temperature limit is given by a step function. The MFT equations are solved iteratively, either synchronously or asynchronously, under annealing in T.

The generic treatment of constraints here is very different from that in a more conventional heuristic treatment of optimization problems. For example in the case of graph bisection one typically starts in a configuration where the nodes are equally partitioned and then proceeds by swapping pairs subject to some acceptance criteria. The constraint of equal partition is respected throughout the updating process. This is in sharp contrast to a neural network technique, where the constraints are implemented in a 'soft' manner by a penalty term. The final MFT solutions could therefore sometimes suffer from a minor imbalance which is easily remedied, either by applying a *greedy heuristic* to the solutions, or by reheating the system and letting it anneal down again.

Very good numerical results were obtained for the graph bisection problem in [12] for problem sizes ranging from 20 to 2000. The quality of the solutions was comparable to those of the CPU-demanding simulated annealing method. The time consumption is lower than any other known method. The method of course becomes even more competitive with respect to time consumption if the intrinsic parallelism is exploited on dedicated hardware.

5.4.3 Mean field dynamics

The dynamics of feed-back ANN typically exhibit a behaviour with two phases (cf. Figure 5.5): at large enough temperatures ($T \to \infty$) the system relaxes into the trivial fixed point $v_i^{(0)} = 0$.

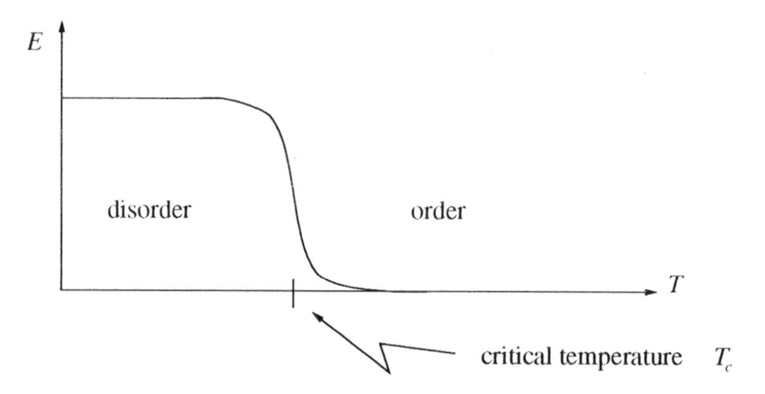

Figure 5.5: E as a function of T illustrating a phase transition

As the temperature is lowered a phase transition is passed at $T = T_c$ and as $T \to 0$ fixed points $v_i^{(*)} = \pm 1$ emerge representing a specific decision made as to the solution to the optimization problem in question (see Figure 5.6). The position of T_c, which here depends

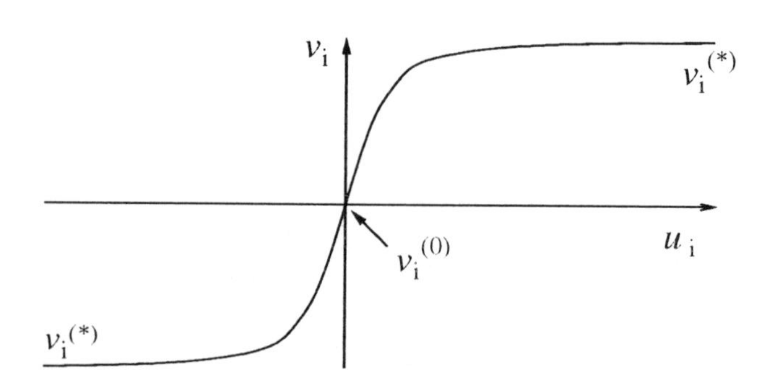

Figure 5.6: Fixed points in $\tanh(u_i)$

on ω_{ij} and α, can be estimated by expanding the sigmoid function (tanh) in a power series around $v_i^{(0)} = 0$ (see Figure 5.6). The fluctuations around $v_i^{(0)}$

$$v_i = v_i^{(0)} + \epsilon_i \tag{5.19}$$

satisfy

$$\epsilon_i = \frac{1}{T}\sum_j m_{ij}\epsilon_j \tag{5.20}$$

where $m_{ij} = \omega_{ij} - \alpha$. For synchronous updating it is clear that if one of the eigenvalues of the matrix $\mathbf{m}/T$ in equation (5.20) is > 1 in absolute value, the fixed point becomes unstable and the solutions will wander away into the nonlinear region. In the case of serial updating the philosophy is the same but the analysis slightly more complicated. We refer the reader to [8] for a more detailed discussion. Finding the largest eigenvalue of $\mathbf{m}$ could be computationally explosive by itself. Is there an approximate way of doing it? Yes, providing $\langle\omega_{ij}\rangle$ and the corresponding standard deviation σ turn out to be sufficient for obtaining estimates within 10% of T_c. Also, this analysis is important for avoiding oscillatory behaviour [14], which appears for eigenvalues < -1. We return to a more general and detailed treatment of this phase structure analysis in section 5.4.

With this method of estimating T_c in advance we can construct a reliable, parallelizable 'black box' algorithm for solving problems of this kind. A prescription is given in section 5.5.

5.5 The Graph Partition Problem

A generalization of the graph bisection problem is *graph partitioning* (GPP), where the N nodes are to be partitioned into K subsets of N/K nodes, again with minimal cutsize (see Figure 5.4b). This requires the introduction of a second index for the neurons. This can be done in two different ways, *neuron multiplexing* and *K-state neurons*. It turns out that the latter have several advantages.

5.5.1 Neuron multiplexing—Ising representation

We define the neurons in this case as follows

$$s_{ia} = 0, 1 \tag{5.21}$$

where the index i denotes the *node* ($i = 1, ..., N$) and a the *subset* ($a = 1, ..., K$). The neuron s_{ia} takes the value 1 or 0 depending on whether node i belongs to set a or not[2]. By analogy with equation (5.7) we choose the energy

$$E = \frac{1}{2}\sum_{ij}\sum_{a\neq b}\omega_{ij}s_{ia}s_{jb} + \frac{\beta}{2}\sum_{i}\sum_{a\neq b}s_{ia}s_{ib} + \frac{\alpha}{2}\sum_{a}(\sum_{i}s_{ia} - \frac{N}{K})^2 \tag{5.22}$$

The first term minimizes the cutsize as in the bisection case and the last term represents the *global constraint* of equipartition; it is zero only if each of the K sets contains N/K nodes. The second term represents an additional *syntax constraint* that penalizes situations where a node ends up in more than one partition—only one a is allowed for each i. Again we define mean field variables, $v_{ia} = \langle s_{ia}\rangle_T$ and the corresponding MFT equations are given by

$$v_{ia} = \frac{1}{2}\left[1 + \tanh\left(-\frac{\partial E[\vec{v}]}{\partial v_{ia}}/T\right)\right] \tag{5.23}$$

The solution space of these MFT equations consists of the interior of the direct product of N K-dimensional hypercubes. In Figure 5.7 we show the cube corresponding to $K = 3$. Next we turn to an alternative encoding which compactifies the solution space by one dimension.

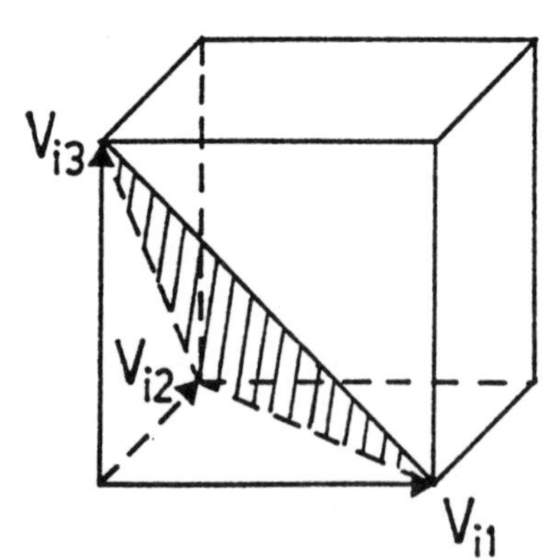

Figure 5.7: The volume of solutions corresponding to the neuron multiplexing encoding for K=3. The shaded plane corresponds to the solution space of the corresponding Potts encoding.

[2]We use $\{0,1\}$ notation here in order to get a more convenient form of the energy function.

5.5.2 K-state neurons—Potts representation

The syntax term will be redundant, if the state space of the neurons is restricted such that exactly one neuron at every site is allowed to be on. This restriction can be compactly written as

$$\sum_a s_{ia} = 1 \tag{5.24}$$

Thus for every i, s_{ia} should be 1 for only one value of a, and 0 for the rest. To explain how this is done, we choose a K-dimensional vector representation of the neuron states: The ath possible state of the neuron vector $\vec{s}_i$ associated with the node i is represented by the unit vector in the a-direction. In this way, the K different states, $(1,0,\ldots,0)$, $(0,1,\ldots,0)$,...,$(0,0,\ldots,1)$, are all normalized and mutually orthogonal. The number of states available at every node is thereby reduced from 2^K to K. Technically, we have a K-state Potts model [15] on our hands. In Figure 5.7 is shown the space of states at one node for the case K=3.

The energy function of equation (5.22) can now be rewritten, using the constraint of equation (5.24), as

$$E = -\frac{1}{2}\sum_{ij}\sum_a \omega_{ij} s_{ia} s_{ja} - \frac{\beta}{2}\sum_i\sum_a s_{ia}^2 + \frac{\alpha}{2}\sum_{ij}\sum_a s_{ia} s_{ja} \tag{5.25}$$

or, in vector notation,

$$E = -\frac{1}{2}\sum_{ij} \omega_{ij} \vec{s}_i \vec{s}_j - \frac{\beta}{2}\sum_i \vec{s}_i^{\,2} + \frac{\alpha}{2}\left(\sum_i \vec{s}_i\right)^2 \tag{5.26}$$

disregarding an unimportant constant term. Note that the second term could now be dropped, since it is a constant. We will however keep it, since for some applications it will improve the solution quality. It also turns out to be a convenient regulator for avoiding chaotic behaviour in synchronous updating (see below). With $\beta = 0$, this expression has exactly the same structure as the energy (equation 5.7) for the graph bisection problem. Indeed, for $K = 2$, they are completely equivalent (apart from a factor 2).

5.5.3 Mean field Potts equations

We next derive the MFT equations corresponding to K-state Potts neurons. Again we start off with the partition function

$$Z = \sum_{[\vec{s}]} e^{-E[\vec{s}]/T} \tag{5.27}$$

Proceeding by analogy with the binary case in the previous section, the partition function is transformed into

$$Z = \int d[\vec{u}]d[\vec{v}] \exp(-E[\vec{v}]/T - \sum_i \vec{u}_i \cdot \vec{v}_i) \sum_{[\vec{s}]} e^{\sum_i \vec{u}_i \cdot \vec{s}_i}. \tag{5.28}$$

Performing the original sum, we are left with

$$Z = \int d[\vec{u}]d[\vec{v}] \exp(-E[\vec{v}]/T - \sum_i \vec{u}_i \cdot \vec{v}_i + \sum_i \log \sum_a e^{u_{ia}}). \tag{5.29}$$

The MFT approximation to $\langle \vec{s}_i \rangle$ is again given by the value of $\vec{v}_i$ at a saddle-point of the effective energy of $[\vec{u}, \vec{v}]$ in the exponent of equation (5.29). Differentiating, we obtain the Potts MFT equations (cf. equations (5.15, 5.16)):

$$u_{ia} = -\frac{\partial E[\vec{v}]}{\partial v_{ia}}/T, \tag{5.30}$$

$$v_{ia} = \frac{e^{u_{ia}}}{\sum_b e^{u_{ib}}}, \tag{5.31}$$

where a denotes the vector components.

From the latter it follows that

$$\sum_a v_{ia} = 1. \tag{5.32}$$

One can thus think of the mean field v_{ia} as the probability for the Potts neuron i to be in state a. The set of values available to $\vec{v}_i$ is the inside of a simplex in a $(K-1)$-dimensional hyperplane, with its corners on the allowed states for the Potts neuron $\vec{s}_i$. For $K = 3$ this is a triangle (cf. Figure 5.7). For $K = 2$, we recover the formalism of the Ising case, provided that we define v_i as $v_{i1} - v_{i2}$, etc. Thus, we have a natural generalization of the MFT approach from $K = 2$ to general multi-state systems.

5.5.4 Mean field dynamics

High temperature

Again we can analyze the linearized equations as in section 4.3 in order to estimate the critical temperature T_c. The value of the critical temperature depends on the self-coupling β, and on precisely how the updating is done. In *serial mode*, neurons are updated one by one, using fresh values of previously updated neurons. In *synchronous mode*, all neurons are updated in parallel, using only old values as input.

At large enough temperatures the system relaxes into the trivial fixed point $v_{ia}^{(o)}$, which is a completely symmetrical state, where all v_{ia} are equal

$$v_{ia}^{(0)} = \frac{1}{K} \tag{5.33}$$

As the temperature is lowered a phase transition is passed at $T = T_c$ and as $T \to 0$ fixed points $v_{ia}^{(*)}$ emerge which are characterized by

$$\Sigma \equiv \frac{1}{N} \sum_{ia} v_{ia}^{(*)2} \to 1 \tag{5.34}$$

where we have introduced an order parameter, the *saturation* Σ. In Figure 5.8 this is illustrated for a $K = 4, N = 100$ graph partition problem. The position of T_c depends on ω_{ij}, α and β. As mentioned above, the fixed point corresponds to the symmetry point of the nonlinear gain functions in the MFT equations (equations (5.30, 5.31)), where these are almost linear (see Figure 5.6). Let us consider fluctuations around the trivial fixed point (cf. equation (5.19))

$$v_{ia} = v_{ia}^{(0)} + \epsilon_{ia} \tag{5.35}$$

First, from equations (5.25, 5.30, 5.31) it follows that

$$\sum_a \epsilon_{ia} = 0 \tag{5.36}$$

so the fluctuations will always be perpendicular to (1,1,1,...). In the linear region the perpendicular components of ϵ_{ia} evolve (suppressing the index a) according to

$$\epsilon_i = \frac{1}{KT} \sum_j M_{ij} \epsilon_j \tag{5.37}$$

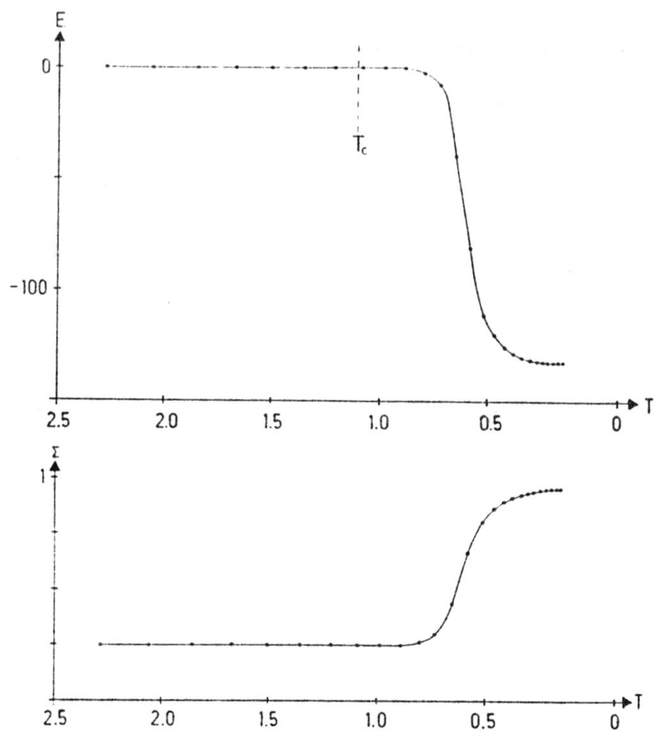

Figure 5.8: (a) Internal energy $E(v_{ia})$ as a function of T for a $K = 4, N = 100$ graph partition problem. Also shown is the prediction from the linearized analysis below for T_c with $\xi = 1.5$. (b) The saturation Σ as a function of T for the same problem.

(cf. equation (5.20)) where

$$M_{ij} = \omega_{ij} - \alpha + \beta\delta_{ij} \tag{5.38}$$

in the case of GPP. For synchronous updating it is clear that if one of the eigenvalues of $\mathbf{M}/KT$ in equation (5.37) is > 1 in absolute value the solutions will wander away into the nonlinear region. Hence T_c will be determined by the eigenvalue distribution of $\mathbf{M}$, and will obviously scale as $1/K$ for a fixed graph. In the case of serial updating the philosophy is the same but the analysis slightly more complicated. We will therefore treat the two cases separately. A more detailed treatment of the contents in this section can be found in [8].

Synchronous Updating In this case T_c is given by

$$T_c = \frac{1}{K}(\textit{largest absolute eigenvalue of } \mathbf{M}) \tag{5.39}$$

It turns out to be convenient to divide $\mathbf{M}$ up into its diagonal and off-diagonal parts

$$M_{ij} = (\beta - \alpha)\delta_{ij} + A_{ij} \tag{5.40}$$

where the off-diagonal part $\mathbf{A}$ is given by

$$A_{ij} = \omega_{ij} - \alpha(1 - \delta_{ij}) \tag{5.41}$$

In terms of the extreme eigenvalues λ_{max} and λ_{min} of $\mathbf{A}$ (which depend on α but *not* β) we thus have

$$T_c = \frac{1}{K}\max[\alpha - \beta - \lambda_{min}, \beta - \alpha + \lambda_{max}] \tag{5.42}$$

We stress that this is an exact equation for T_c. Also note the simple β-dependence; for a fixed α, $T_c(\beta)$ can be obtained by probing it for two values of β (cf. Figure 5.9a). In [8] it is shown, using the average $t = \langle\omega_{ij}\rangle$ and the standard deviation σ_t of the off-diagonal elements of ω_{ij}, that for reasonable values of α, λ_{min} is well approximated by

$$\lambda_{min} = -(N-1)(\alpha - t) \tag{5.43}$$

and that the other eigenvalues are comparatively closely distributed around a mean value $\alpha - t$ with the standard deviation $\sqrt{N}\sigma_t$. We can thus write λ_{max} as

$$\lambda_{max} = \alpha - t + \xi\sqrt{N}\sigma_t \tag{5.44}$$

where $\xi > 0$ is a numerical factor that gives the deviation of λ_{max} from the average in units of the standard deviation. Empirically we find ξ to lie in the region 1.5 to 2.0 for the graph partition problem.

From equation (5.42) we can identify two distinct regions in the (KT, β)-plane, with different behaviours (see Figure 5.9a).

Region (1) In this region one has

$$\beta \leq \frac{N}{2}\alpha - \frac{N-2}{2}t - \frac{1}{2}\xi\sqrt{N}\sigma_t \tag{5.45}$$

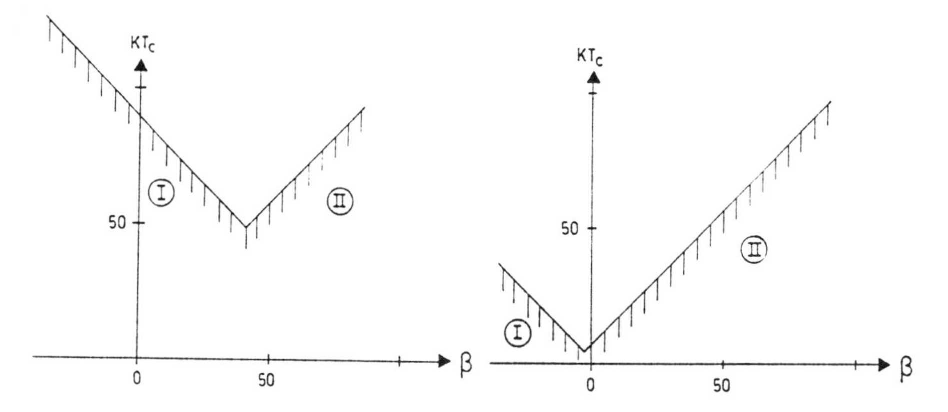

Figure 5.9: (a) The (KT, β)-plane for synchronous updating. The two lines correspond to eigenvalues -1 and $+1$ respectively for $t = 0$. (b) The (KT, β)-plane for serial updating. The two lines correspond to unitary eigenvalues $\neq 1$ and $= 1$ respectively for $t = 0$.

and

$$T_c = \frac{1}{K}[\alpha - \beta + (N-1)(\alpha - t)] \tag{5.46}$$

corresponding to a non-degenerate eigenvalue, value -1, of the evolution matrix in equation (5.40). This is a single eigenvalue. Due to its sign, the trivial fixed point solution will bifurcate into an alternating phase when the temperature is lowered through T_c; all the $\vec{v}_i$ will jump more or less uniformly back and forth. Thus when executing the algorithm synchronously one should avoid this region. Such oscillating behaviour has been observed in spin-glass systems when updating MFT equations synchronously [14].

Region (2) In this region one has

$$\beta \geq \frac{N}{2}\alpha - \frac{N-2}{2}t - \frac{1}{2}\xi\sqrt{N}\sigma_t \tag{5.47}$$

and

$$T_c = \frac{1}{K}[\beta - t + \xi\sqrt{N}\sigma_t] \tag{5.48}$$

corresponding to the eigenvalue $+1$, which in the limit of small fluctuations (σ_t) is $(N-1)$-fold degenerate. Hence the situation is very different. The dynamics will be smooth with no oscillations. Owing to the high approximate degeneracy of the relevant eigenvalue, the time evolution of the system will have a chance to 'feel' its way to a good solution.

Serial Updating The analysis for this case is somewhat more elaborate. Again the relevant matrix is $\mathbf{M}/KT$, but in this case the updating proceeds row by row, and the linearized updating equation for the perpendicular fluctuations ϵ_{ia} (cf. equation(5.37)) now reads

$$\epsilon'_i = \frac{1}{KT}(\sum_{j=1}^{i-1} M_{ij}\epsilon'_j + \sum_{j=i}^{N} M_{ij}\epsilon_j) \tag{5.49}$$

where new values ϵ'_j are used for the previously updated components. Dividing $\mathbf{M}$ up into an upper ($\mathbf{U}$), a diagonal ($\mathbf{D}$) and a lower ($\mathbf{L}$) part, equation(5.49) can be rewritten in matrix notation

$$\epsilon' = \frac{1}{KT}[\mathbf{L}\epsilon' + (\mathbf{D} + \mathbf{U})\epsilon]\ . \tag{5.50}$$

Rearranging, we obtain

$$\epsilon' = (KT - \mathbf{L})^{-1}(\mathbf{U} + \mathbf{D})\epsilon \tag{5.51}$$

where the LHS is the effective synchronous updating matrix that corresponds to serial updating. In other words we have 'synchronized' the analysis and again T_c is characterized by a unit modulus for the largest eigenvalue μ. In the case of $\mu = 1$, the result is the same as that of synchronous updating. For $|\mu| = 1, \mu \neq 1$, it follows from a Hermiticity argument that this is possible only for $T = (1/K)(\alpha - \beta)$. Thus one obtains for the critical temperature

$$T_c = \frac{1}{K}\max[\alpha - \beta, \beta - \alpha + \lambda_{max}] \rightarrow T_c = \frac{1}{K}\max[\alpha - \beta, \beta - t + \xi\sqrt{N}\sigma_t] \tag{5.52}$$

with ξ as in equation (5.44), and again we find two regimes (see Figure 5.9b).

Low temperature

At low temperature, any difference between the components of a neuron's input will be strongly magnified. As a result, we have a *winner-take-all* situation, where one neuron component will be almost one, and the others close to zero. The dynamics become discrete, and a kind of local optimization results as $T \rightarrow 0$. Eventually the network should settle at a fixed point close to an allowed state of the Potts spin system, representing a (possibly approximate) solution to the optimization problem.

The β value also has an impact here: a higher value tends to stabilize bad decisions (positive feed-back), while a lower value destabilizes even a good solution, and can lead to cyclic or even chaotic behaviour. For serial mode, it is never a problem to choose a β that is satisfactory in both temperature limits [8]. This is often not the case for synchronous mode, so serial mode will be understood where not otherwise stated.

5.5.5 A generic algorithm

For most problems, the proper values of β and T_0 can be calculated or estimated in advance. The complete neural network algorithm for a generic K-state Potts system will then look as follows:

A generic Potts neural network algorithm

1. Choose problem - ω_{ij}
2. Find the approximate phase transition temperature by linearizing equation (5.31). (for details see [8])
3. Add a β-term if necessary.
4. Initialize the neurons v_{ia} with 1/K $\pm$ random values, and set $T_0 = T_c$.
5. Until $\Sigma \equiv \frac{1}{N} \sum_{ia} v_{ia} \geq 0.99$, do:
   - 5.1 At each T_n update all v_{ia}; $v_{ia} = e^{u_{ia}} / \sum_b e^{u_{ib}}$.
   - 5.2 $T_{n+1} = 0.9 \times T_n$.
6. After $\Sigma \geq 0.99$ is reached perform a greedy heuristic if needed to account for possible imbalances or rule violations.

(The algorithm is not very sensitive to the choice of Σ (5), and annealing rate (5.2). The quoted values of 0.99 and 0.9 were found to be optimal for the applications studied here.)

5.5.6 Numerical results

In [8] the neural approach to graph partition was compared with that of simulated annealing. Not unexpectedly it turns out that the neuron multiplexing method requires 'custom-made' tailoring of parameters for each application. This is not acceptable. We interpret this weakness as being due to the redundancy in the encoding—the dynamics is not explicitly confined to the hyperplane ($\sum_a v_{ia} = 1$). The Potts encoding, on the other hand, turns out to be much more stable. We therefore stick to this approach in what follows. In Figure 5.10 we compare the performance of the Potts neural network with simulated annealing. The graphs were generated by randomly connecting the nodes with probability $P = 10/N$. The results are impressive! The neural network algorithm performs as well as (in some cases even better than) the simulated annealing method with its excessive annealing and sweep schedule (see [8]). This is accomplished with a very modest number of iterations, $\mathcal{O}(50-100)$. In fact the number of iterations is empirically independent of problem size [8].

5.6 The Travelling Salesman Problem

In the travelling salesman problem (TSP) a closed tour with minimal distance is to be chosen such that each city is visited exactly once. To encode this we define neurons s_{ia} to be 1 if city i has the ath tour number and 0 otherwise, and let d_{ij} be the distance between city i and j. This problem (see Figure 5.4c) is somewhat related to graph partition and based on our experience above we stick to the Potts representation, equation (5.24), in which the energy reads

$$E = \sum_{ij=1}^{N} d_{ij} \sum_{a=1}^{K} s_{ia} s_{j(a+1)} - \frac{\beta}{2} \sum_{i=1}^{N} \sum_{a=1}^{K} s_{ia}^2 + \frac{\alpha}{2} \sum_{a} (\sum_{i} s_{ia})^2 \quad (5.53)$$

where we again have included an auxiliary β-term for dynamical reasons[3]. As in the graph partition case the energy is minimized by iteratively solving the Potts MFT equations (equations (5.30, 5.31)) using the generic prescription of Section 5.5.5. Again an analysis of

[3]This is of course not the only way to code the problem. We could e.g. interchange the roles of the labels i and a.

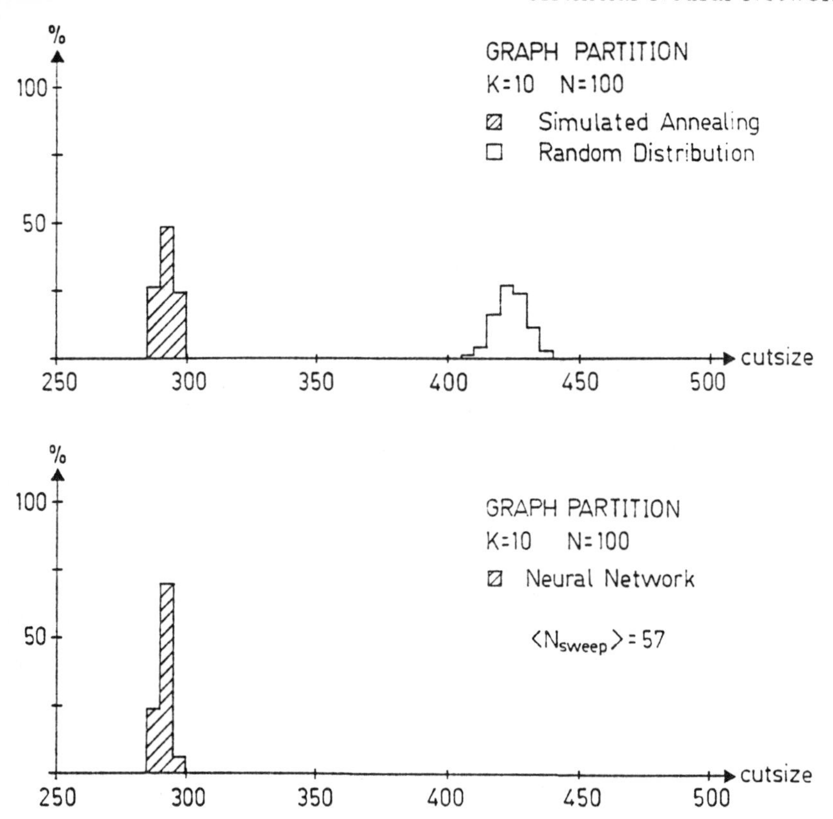

Figure 5.10: Comparison of Potts neural network solutions versus simulated annealing and random distributions for a 10×100 graph partition problem. The histograms are based on 50 experiments for the neural network and simulated annealing algorithms and 1000 for the random distributions (from [8]).

the linearized dynamics along the lines of the previous sections can be used to estimate T_c. In [8] the Potts approach for the TSP was explored numerically for problem sizes up to $N_{cities}=200$, again with encouraging results. The neural network solution qualities fall on the tail of the simulated annealing ones. Defining the relative quality q as the ratio

$$q = \frac{\langle E(\text{random config.})\rangle - \langle E(\text{neur. netw.})\rangle}{\langle E(\text{random config.})\rangle - \langle E(\text{sim. ann.})\rangle} \tag{5.54}$$

we find that q never falls below 92%. Furthermore, as in the graph partition case, there are no really bad solutions.

As far as convergence time goes, we have observed no K or N dependence for the problem sizes examined. The comparisons with simulated annealing concern quality only. When discussing the total time consumption one has to distinguish between serial and parallel implementations. With serial execution, the time consumption τ of the Potts neuron algorithm is $\propto N^3$ for TSP, as in the neuron multiplexing case [7]. This should be compared with N^2 for simulated annealing. The real advantage lies in the inherent parallelism, which when exploited gives $\tau \propto$ *constant* for a general purpose parallel machine or custom made hardware.

For TSP there is also an alternative neural approach, the deformable templates, to which we will return in section 8.

5.7 Scheduling Problems

A Potts neural network formulation is almost ideal for scheduling problems—these have a natural formulation in terms of Potts systems. In its purest form, a scheduling problem consists entirely in fulfilling a set of basic constraints, each of which can be encoded as a penalty term that will vanish when the constraint is obeyed. Thus, the minimum energy is known (= 0), and one can identify exact (legal) solutions by inspection. In many applications, there exist additional preferences within the set of legal schedules.

First we will discuss a synthetic problem [16], where the principles of the neural mapping are very transparent. Then we will briefly discuss how to deal with the additional complication in a real-world problem [17], in this case a Swedish high school.

5.7.1 A synthetic example

In [16] a simplified scheduling problem, where N_p teachers give N_q classes in N_x class rooms at N_t time slots, was mapped onto a Potts neural network. In this problem one wants solutions where all N_p teachers give a lecture to each of the N_q classes, using the available space-time slots with no conflicts in space (class rooms) or time. These are the *basic constraints* that have to be satisfied. In addition various *preferences* regarding continuity in class-rooms etc. were considered. The basic entities of this problem can be represented by four sets consisting of N_p,N_q,N_x and N_t elements respectively.

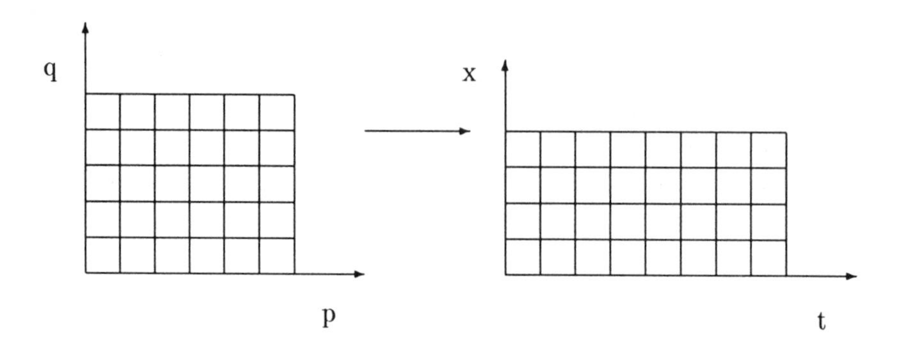

Figure 5.11: Mapping of events (p, q) onto space-time slots (x, t)

There is a very transparent way to describe this problem that naturally lends itself to the Potts neural encoding, where *events*, defined by teacher-class pairs (p, q), are mapped onto *space-time slots* (x, t) (see Figure 5.11). Potts neurons $s_{pq;xt}$ are defined to be 1 if the event (p, q) takes place in the space-time slot (x, t) and 0 otherwise. The basic constraints in this picture are as follows:

1. an event (p, q) should occupy precisely one space-time slot (x, t);
2. different events (p_1, q_1) and (p_2, q_2) should not occupy the same space-time slot (x, t);
3. a teacher p should have at most one class at a time;
4. a class q should have at most one teacher at a time.

A schedule fulfilling all the basic constraints is said to be *legal*. The first constraint can be embedded in the neural network in terms of the Potts normalization condition

$$\sum_{x,t} s_{pq;xt} = 1 \tag{5.55}$$

for each event (p, q). In other words we have $N_p N_q$ neurons, each of which has $N_x N_t$ possible states. The other three constraints are implemented using energy penalty terms as follows:

$$E_{XT} = \frac{1}{2} \sum_{x,t} \sum_{p_1,q_1} \sum_{p_2,q_2} s_{p_1 q_1;xt} s_{p_2 q_2;xt} = \frac{1}{2} \sum_{x,t} [\sum_{pq} s_{pq;xt}]^2 \tag{5.56}$$

$$E_{PT} = \frac{1}{2} \sum_{p,t} \sum_{q_1,x_1} \sum_{q_2,x_2} s_{pq_1;x_1t} s_{pq_2;x_2t} = \frac{1}{2} \sum_{pt} [\sum_{qx} s_{pq;xt}]^2 \tag{5.57}$$

$$E_{QT} = \frac{1}{2} \sum_{q,t} \sum_{p_1,x_1} \sum_{p_2,x_2} s_{p_1q;x_1t} s_{p_2q;x_2t} = \frac{1}{2} \sum_{qt} [\sum_{px} s_{pq;xt}]^2 \tag{5.58}$$

This way of implementing the constraints is by no means unique. In particular one can add various diagonal terms that merely have the effect of adding a fixed constant to the energy. Again mean field variables are introduced, $v_{pq;xt} \sim \langle s_{pq;xt} \rangle_T$ and the corresponding MFT equations read (cf. equations (5.30, 5.31))

$$u_{pq;xt} = -\frac{1}{T} \cdot \frac{\partial E}{\partial v_{pq;xt}} \tag{5.59}$$

$$v_{pq;xt} = \frac{e^{u_{pq;xt}}}{\sum_{x't'} e^{u_{pq;x't'}}}. \tag{5.60}$$

This is a straightforward approach that works well. However, an important simplification can be made. It turns out that with the encoding $v_{pq;xt}$ the MFT-equations give rise to two phase transitions; one in x and one in t. In other words the system naturally factorizes into two parts. It is therefore economical to implement this factorization already at the encoding level. This is done by replacing $s_{pq;xt}$ by x-neurons $s^{(X)}_{pq;x}$ and t-neurons $s^{(T)}_{pq;t}$

$$s_{pq;xt} = s^{(X)}_{pq;x} s^{(T)}_{pq;t} \tag{5.61}$$

with separate Potts conditions replacing equation (5.55)

$$\sum_{x} s^{(X)}_{pq;x} = 1 \tag{5.62}$$

and

$$\sum_{t} s^{(T)}_{pq;t} = 1 \tag{5.63}$$

respectively. This means that the number of degrees of freedom reduces from $N_p N_q N_x N_t$ to $N_p N_q (N_x + N_t)$. As a consequence the sequential execution time goes down.

The performance of this algorithm was investigated [16, 17] for a variety of problem sizes and for different levels of difficulty measured by the ratio between the number of events and the number of available space-time slots. It was found that the algorithm consistently found legal solutions with very modest convergence times for problem sizes $(N_p, N_q) = (5,5),\ldots,(12,12)$. With convergence time we here mean the total number of sweeps needed to obtain a legal solution no matter how many trials it takes. Also when introducing preferences, e.g. for having subsequent lessons in the same room, very good solutions were obtained.

5.7.2 A realistic example

The synthetic scheduling problem of [16] contains several simplifications as compared to realistic problems. In [17] more realistic problems from the Swedish high school system were explored, which required an extended formalism. Here we give a list of items that an extended formalism will need to handle.

1. One week periodicity (occasionally extended to two- or four-week periodicity).

2. In [16] each teacher has a class exactly once. In this case he has to give lessons in certain subjects a few hours a week, appropriately spread out.

3. In [16] it was assumed that all class rooms were available for all subjects. This is not the case in reality. Many subjects require special purpose rooms.

4. Many subjects are taught for two hours in a row (double hours).

5. Group formation. For some optional subjects the classes are broken up into option groups temporarily forming new classes.

6. Various preferences have to be considered.

When taking these points into account in our formalism we cannot keep p and q as the independent quantities. Rather we define an independent variable i (event index) to which p and q are attributes, $p(i)$ and $q(i)$. The values of these exist in a look-up table. This table contains all the necessary information to process each event i. The time index t we subdivide into weekdays (d) and daily hours (h). The Potts conditions (equations (5.62, 5.63)) now read

$$\sum_x s_{i;x}^{(X)} = 1 \tag{5.64}$$

and

$$\sum_t s_{i;t}^{(T)} = 1 \tag{5.65}$$

This also implies that the Potts neurons will have a different number of components. This situation can be handled when analyzing the MFT dynamics in terms of fixed points and T_c [17].

To facilitate the handling of double hours we introduce effective t-neurons $\tilde{s}_{i;t}^{(T)}$ defined as

$$\tilde{s}_{i;t}^{(T)} = \sum_{k=0}^{g_i-1} s_{i;t-k}^{(T)} \tag{5.66}$$

where the multiplicity g_i is *1* for single hours and *2* for double hours.

There are basically four kinds of *preferences* we encounter when scheduling Swedish high schools:

1. the lunch 'lesson' has to appear within 3 hours around noon;
2. the different lessons for a class in a particular subject should be *spread* as much as possible over the week days;
3. the schedules should have as few 'holes' as possible: lessons should be *glued* together;
4. teachers may have various individual preferences.

The first of these points is really a syntactic constraint rather than a preference, easily taken into account by restricting the Potts index range for the relevant events. Spreading, gluing and teacher preferences can be accommodated by an appropriate choice of penalty functions (see [17]).

As in the case of graph problems and TSP above, an automated implementation of the algorithm exists along the lines of the generic prescription defined in section 5.5. In Figure 5.12 we show a typical evolution of E_{hard} (the basic constraint part), Σ_T, Σ_X and T as a function of N_{sweep} for a Swedish high school problem. These problems typically have $\sim$ 90 teachers, $\sim$ 50 weekly hours, $\sim$ 45 classes and $\sim$ 60 class-rooms, which corresponds to $\sim 10^{4600}$ possible choices. In the factorized Potts formulation these are handled by $\sim 10^5$ neural degrees of freedom. High quality solutions emerge. A revision capability is inherent in our formalism which is useful when encountering unplanned events once a schedule exists. Such rescheduling is performed by clamping those events not subject to change and heating up and cooling the remaining dynamical neurons.

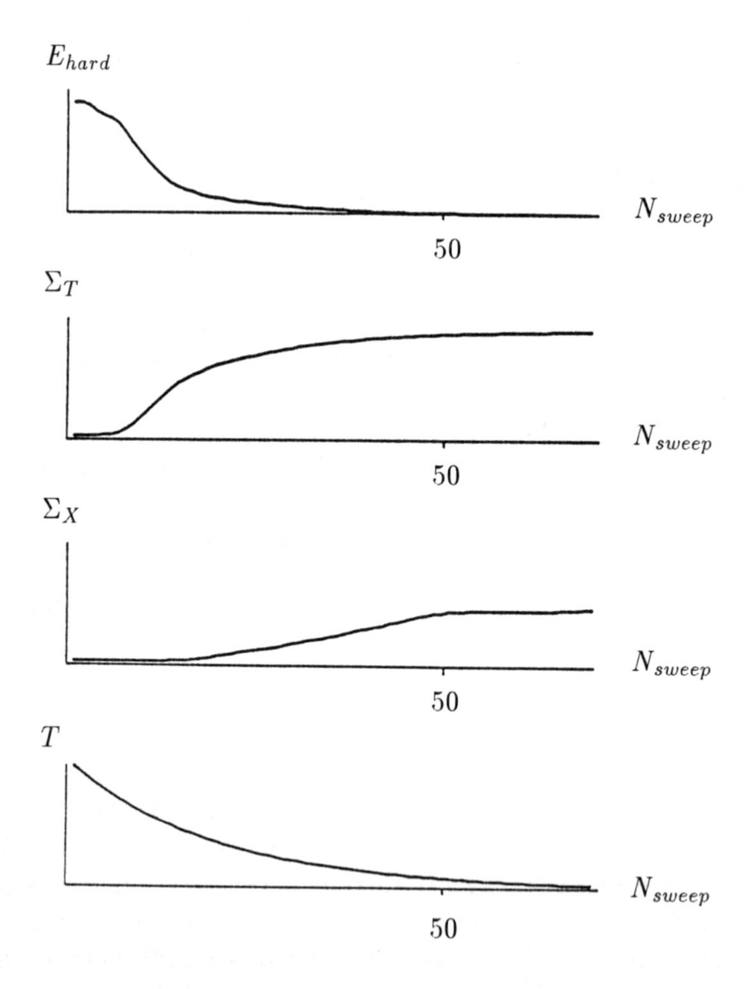

Figure 5.12: Energy (E_{hard}), saturations (Σ_T, Σ_X) and temperature (T) as functions of N_{sweep} for one run with a Swedish high school problem

One should keep in mind that problems of this kind and size are so complex that even several man-months of human planning will in general not yield solutions that will meet all the requirements in an optimal way. We have not been able to find any algorithm in the literature that solves a real-world problem with this complexity. Existing commercial software packages do not solve the entire problem. Rather the problem is solved with an interactive user taking step-wise decisions.

The scheduling problems we considered here were all based upon equality constraints. There are many resource allocation problems which contain inequalities. In Section 5.9 we will show how to deal with these within the ANN framework.

5.8 Deformable Templates

For low-dimensional geometric optimization problems it is sometimes advantageous to abandon the pure Potts probabilistic description in favour of an expressly geometric template formulation, thereby reducing the number of degrees of freedom. For the travelling salesman problem, such a formulation is given by the *elastic net* approach [9], to be described below.

Denote the N city positions in the TSP by $\vec{x}_i$ (see Figure 5.13). We are going to match these cities with $M > N$ template coordinates $\vec{y}_a$ such that $\sum_a |\vec{y}_a - \vec{y}_{a+1}|^2$ is minimized and each $\vec{x}_i$ is matched by precisely one $\vec{y}_a$. For each city i define a Potts spin s_{ia} to be 1 if a is matched to i and 0 otherwise. Consider the following energy expression, to be minimized w.r.t. s_{ia} and $\vec{y}_a$:

$$E(s_{ia}, \vec{y}_a) = \frac{1}{2T} \sum_{ia} s_{ia} |\vec{x}_i - \vec{y}_a|^2 + \frac{\gamma}{2} \sum_a |\vec{y}_a - \vec{y}_{a+1}|^2 \qquad (5.67)$$

The first term enforces matching: it is minimized when each $\vec{x}_i$ coincides with that $\vec{y}_a$ for which $s_{ia} = 1$. The second term is a kind of tour length term. As $T \to 0$, the full E will be minimized when the $\vec{y}_a$ equidistantly trace out the shortest closed tour visiting all cities.

The parameter γ governs the relative strength between matching and tour length (cf. α and β in the neural descriptions above), and should be suitably chosen depending on M, N and the typical distances in the problem.

The problem is thus encoded in terms of two distinct types of variables: the Potts spins s_{ia}, and the template coordinates $\vec{y}_a$. The Boltzmann distribution for the energy in equation (5.67) reads

$$P(s_{ia}, \vec{y}_a; T) = \frac{e^{-E(s_{ia}, \vec{y}_a)/T}}{Z} \tag{5.68}$$

with the partition function Z given by

$$Z = \sum_{s_{ia}} \int d\vec{y}_a e^{-E(s_{ia}, \vec{y}_a)/T} \tag{5.69}$$

We can now define so-called *marginal distributions* by integrating out either the Potts spin or the template coordinate degrees of freedom. If we choose the latter alternative we end up with a pure neural description of the problem similar to the one in the previous section. We will here choose the former alternative. Performing the sum over s_{ia} [18, 19] we obtain

$$Z = \int d\vec{y}_a e^{-\gamma \sum_a |\vec{y}_a - \vec{y}_{a+1}|^2/2T} \prod_i \left(\sum_a e^{-|\vec{x}_i - \vec{y}_a|^2/2T^2} \right) \tag{5.70}$$

The summation is over only those configurations where s_{ia} is 1 for only one a for each i. We rewrite equation (5.70) as

$$Z = \int d\vec{y}_a e^{-E_{eff}(\vec{y}_a)/T} \tag{5.71}$$

where the *effective energy* E_{eff} is given by

$$E_{eff}(\vec{y}_a) = -T \sum_i \log(\sum_a e^{-|\vec{x}_i - \vec{y}_a|^2/2T^2}) + \gamma \sum_a |\vec{y}_a - \vec{y}_{a+1}|^2/2 \tag{5.72}$$

Next we minimize E_{eff} with respect to $\vec{y}_a$ using gradient descent

$$\Delta \vec{y}_a = \eta \left[\sum_{ia} \tilde{v}_{ia}(\vec{x}_i - \vec{y}_a)/T + \gamma(\vec{y}_{a+1} - 2\vec{y}_a + \vec{y}_{a-1}) \right] \tag{5.73}$$

where η is a time-step, and the Potts factor (cf. equation (5.31)) $\tilde{v}_{ia}$ is given by

$$\tilde{v}_{ia} = \frac{e^{-|\vec{x}_i - \vec{y}_a|^2/2T^2}}{\sum_b e^{-|\vec{x}_i - \vec{y}_b|^2/2T^2}} \tag{5.74}$$

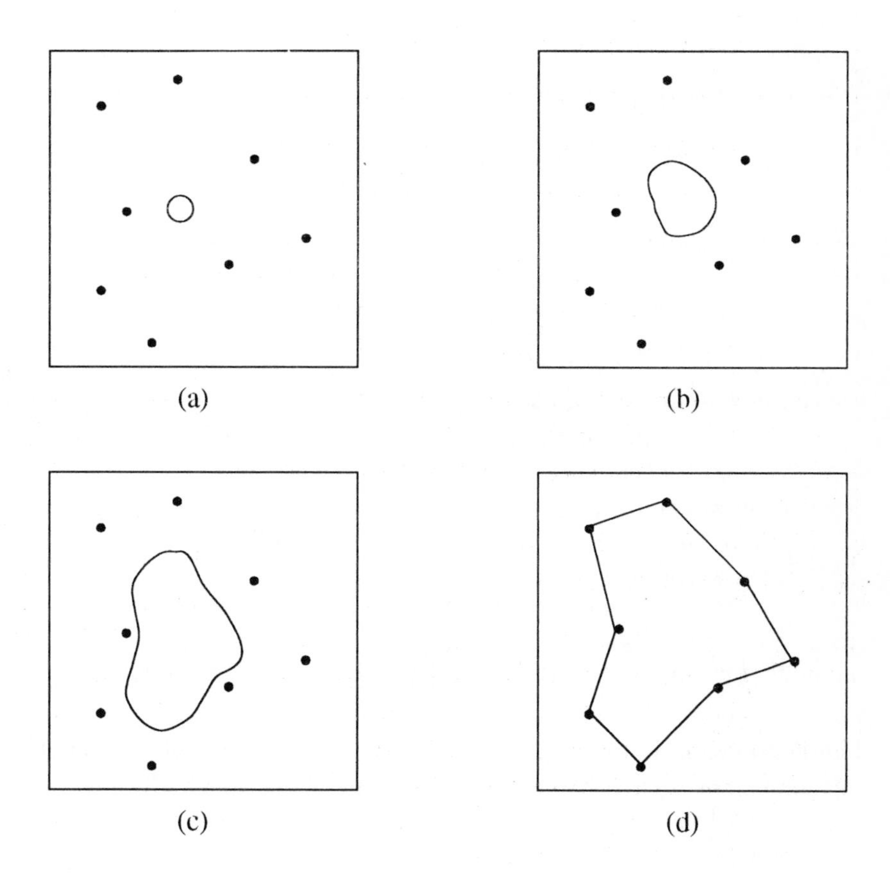

Figure 5.13: Evolution of the template 'snake' from high to low temperatures

A typical snake evolution is schematically depicted in Figure 5.13. Here, in contrast to the Potts description in the previous section, the logical units are only implicit in the dynamics; the primary dynamical degrees of freedom are instead the analog variables $\vec{y}_a$. The full algorithm will look as follows.

An elastic net algorithm for the TSP

1. Choose problem: $\{\vec{x}_i,\, i = 1, N\}$.
2. Choose number of templates M, $M \gg N$.
3. Choose suitable γ and initial temperature T_0.
4. Compute the centre of gravity of the cities $\vec{x}_i$ and displace it slightly with a random seed. Place templates $\vec{y}_a$ with equal spacing on a circle around this centre.
5. Until the system has settled, do:

   5.1 Update templates $\vec{y}_a$;
   $\vec{y}_a(T_n) = \vec{y}_a(T_{n-1}) + \Delta\vec{y}_a$
   where $\Delta\vec{y}_a$ is given by equation (5.73).

   5.2 $T_n = 0.9 \times T_{n-1}$.

How does this algorithm work? The first term in equation (5.72) contains a sum of Gaussians around the templates with width T. At high T this makes all the $\vec{y}_a$ compete to match all cities; in effect they will be attracted towards the centre of mass. As T decreases, the Potts factors become more selective, and the range of competition for each $\vec{x}_i$ is focused on a smaller neighbourhood. Finally one $\vec{y}_a$ is singled out to match each $\vec{x}_i$, while the remaining $\vec{y}$s become arranged equidistantly along straight lines connecting the cities.

This algorithm produces very high quality solutions [13]. And, very importantly, an N-city problem requires only $2N$ degrees of freedom. For problems like the TSP, embedded in a low dimensional space, a template method is to be preferred to the pure neural (Potts) one. However, for high-dimensional or non-geometric problems, such as scheduling [16, 17], the Potts neural network approach is ideal.

The structure of the first term in equation (5.73) is similar to the one for collective updating of self-organizing networks [20, 21]. The templates $\vec{y}_a$ play the role of Gaussian centres that adapt to the distribution of cities. The main differences between this approach and the self-organizing networks are

1. the neighbourhood function, here given by $\Lambda = \tilde{v}_{ia}$, i.e. each 'weight' (template) is updated in some sense through the Potts updating equation (equation (5.74));

2. the number of templates, which is here $\gg N$.

This deformable templates approach has also been successfully applied to track finding [19, 22, 23]

5.9 The Knapsack Problem

The application areas dealt with above (travelling salesman, graph partition and scheduling) are characterized by having low-order polynomial *equality* constraints. Hence they can be implemented by polynomial penalty terms. However, in many other optimization problems, in particular those of resource allocation type, one has to deal with *inequalities*. The objective of this section is to develop a mapping and MFT method to deal with this kind of problem. As a typical resource allocation problem we choose the 0-1 knapsack problem for our studies.

In the knapsack problem (as described in chapter 1) we have a set of N *items*, where item i has *utility* c_i and *load* a_{ki}. The goal is to fill a 'knapsack' with a subset of the items such that their total utility,

$$U = \sum_{i=1}^{N} c_i s_i \tag{5.75}$$

is maximized, subject to a set of M load constraints

$$\sum_{i=1}^{N} a_{ki} s_i \leq b_k; \qquad k = 1, \ldots, M \tag{5.76}$$

defined by load *capacities* b_k.

In equations (5.75, 5.76) s_i are binary $\{0, 1\}$ decision variables, representing whether item i goes into the knapsack. The variables (c_i, a_{ki} and b_k) that define the problem are all real numbers.

We will consider a class of problems, where a_{ki} and c_i are independent uniform random numbers on the unit interval, while b_k are fixed to a common value b. With a b above $N/2$, the problem becomes trivial—the solution will have almost all $s_i = 1$. Conversely, with $b \ll N/4$, the number of allowed configurations will be small and an exact solution can easily be found. We pick the most difficult case, defined by $b = N/2$. The expected number of used items in an optimal solution will then be about $N/2$, and an exact solution becomes inaccessible for large N.

In the optimal solution to such a problem, there will be a strong correlation between the value of c_i and the probability for s_i to be 1. With a simple heuristic based on this observation, one can often obtain near-optimal solutions very fast. We will therefore also consider a class of harder problems with more homogeneous c_i distributions—*homogeneous* problems. The extreme case is when c_i are constant, and the utility proportional to the number of items used.

We note in passing that the *set covering* problem is a special case of the general problem, with random $a_{ki} \in \{0,1\}$, and $b_k = 1$. This defines a comparatively simple problem class, according to the above discussion, and we will stick to the knapsack problem in what follows.

We start by mapping the problem defined in equations (5.75, 5.76) onto a generic neural network energy function E,

$$E = -\sum_{i=1}^{N} c_i s_i + \alpha \sum_{k=1}^{M} \Phi\left(\sum_{i=1}^{N} a_{ki} s_i - b_k\right) , \tag{5.77}$$

where Φ is a penalty function to ensure that the constraint in equation (5.76) is fulfilled. The coefficient α governs the relative strength between the utility and constraint terms. For equality constraints an appropriate choice of $\Phi(x)$ would be $\Phi(x) = x^2$. Having inequalities we need a $\Phi(x)$ that only penalizes configurations for which $x \geq 0$. One possibility is to use a sigmoid, $\Phi(x) = g(x;T)$ (see Figure 5.14a). This option has the potential disadvantage that the penalty becomes constant ($= \alpha$) for large constraint violations. An alternative that gives penalty in proportion to the degree of violation is

$$\Phi(x) = x\Theta(x) \tag{5.78}$$

This function (see Figure 5.14b) has the additional advantage that no extra parameter like the temperature T in the sigmoid is needed. The slope of Φ is implicitly given by α in equation (5.77). The $x\Theta(x)$ alternative consistently gives better performance and is used in what follows.

Minimizing equation (5.77) is again done with the mean field approximation. Owing to the non-polynomial form of the constraint in equation (5.78), special care is needed when implementing the MFT

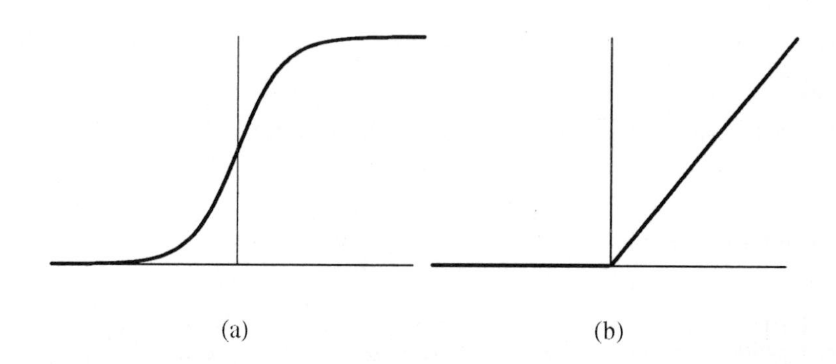

Figure 5.14: (a) A sigmoid $g(x;T)$ (cf. equation 5.2) (b) The penalty function $x\Theta(x)$ of equation (5.78)

approximation. As is shown in [24] this can be done by replacing the derivative $\frac{\partial E}{\partial v_i}$ in

$$v_i = g\left(-\frac{\partial E}{\partial v_i};T\right) \tag{5.79}$$

by a difference,

$$\frac{\partial E}{\partial v_i} \to -c_i+\alpha \sum_{k=1}^{M} \left[\Phi(\sum_{j=1}^{N} a_{kj}v_j - b_k)|_{v_i=1} - \Phi(\sum_{j=1}^{N} a_{kj}v_j - b_k)|_{v_i=0} \right] . \tag{5.80}$$

Equations (5.79, 5.80) are solved iteratively by annealing in T. Again, a more or less automatic scheme for doing this exists (cf. section 5.5.5). The high-T_c fixed point analysis is somewhat more difficult in this case. We refer the reader to [24] for a discussion of this point.

In [24] this ANN approach is compared with other approaches. For reasonably small problem sizes it is feasible to use an exact algorithm, branch and bound, for comparison. For larger problem sizes, one is confined to other approximate methods, such as simulated annealing [10], greedy heuristics and linear programming based on the simplex method (see for example [25]). With the branch-and-bound (BB) tree search technique we can reduce the number of computational steps. This method consists in going down a search tree and checking bounds on constraints or utility for entire subtrees, thereby avoiding unnecessary searching. For non-homogeneous problems in

Table 5.2: Comparison of performance and CPU time consumption for the different algorithms on a N=M=30 problem. The CPU consumption refers to a DEC3100 workstation.

Algorithm	c_i=rand[0,1]		c_i=rand[0.45,0.55]		c_i=0.5	
	Perf.	CPU time	Perf.	CPU time	Perf.	CPU time
BB	1	16	1	1500	1	1500
NN	0.98	0.80	0.95	0.70	0.97	0.75
SA	0.98	0.80	0.95	0.80	0.96	0.80
LP	0.98	0.10	0.93	0.25	0.93	0.30
GH	0.97	0.02	0.88	0.02	0.85	0.02

particular, this method is accelerated by ordering the c_i values according to magnitude:

$$c_1 > c_2 > .... > c_N \ . \tag{5.81}$$

For problems where the constraints are 'narrow' (b not too large) this method can require substantially lower computation needs. However, it is still based on exploration and it is only feasible for problem sizes less than $M = N \approx 30 - 40$.

A greedy heuristic (GH) is a simple and fast approximate method for a non-homogeneous problem. Proceeding from larger to smaller c_i (cf. equation (5.81)), collect every item that does not violate any constraint. This method scales as NM.

Simulated annealing (SA) [10] is easily implemented in terms of attempted single-spin flips, subject to the constraints. Suitable annealing rates and other parameters are given in [24]. This method also scales as NM times the number of iterations needed for thermalization and cooling.

Linear programming (LP) based on the simplex method [25] is not designed to solve discrete problems like the knapsack one. It does apply, however, to a modified problem with $s_i \in [0, 1]$. For the ordered (equation (5.81)) non-homogeneous knapsack problem this gives solutions with a set of leading 1's and a set of trailing 0's, and a window in between containing real numbers. Augmented by greedy heuristics for the elements in this window, fairly good solutions emerge. The simplex method scales as N^2M^2. First we compare the

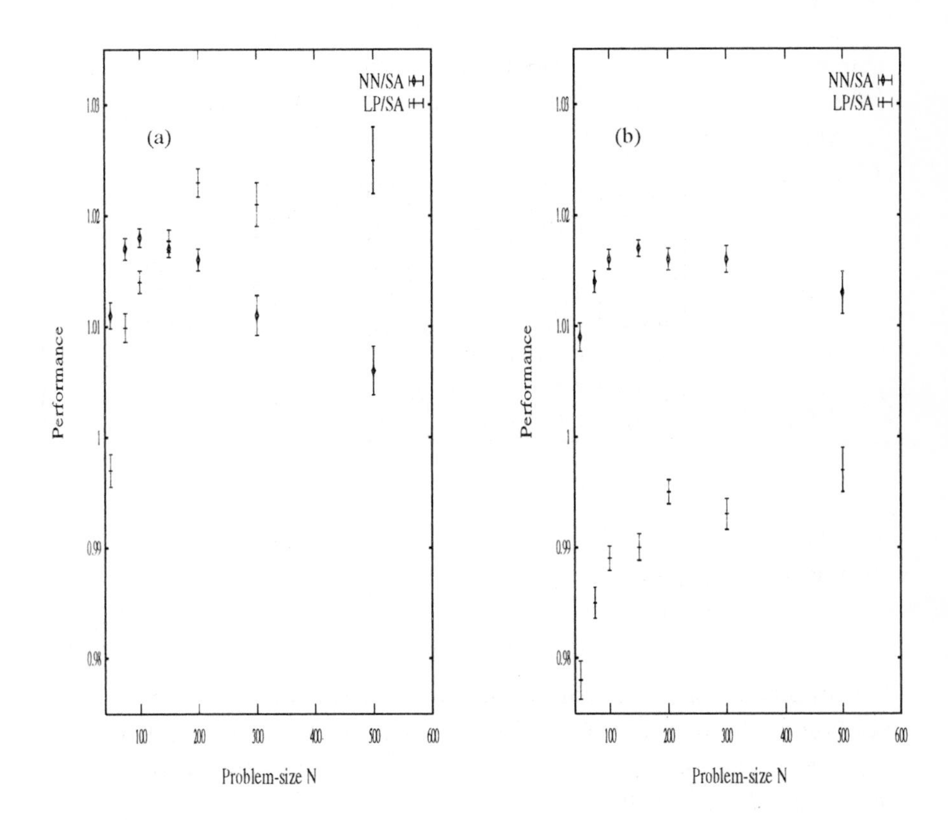

Figure 5.15: Performance of the neural network (NN) and linear programming approaches (LP) normalized to simulated annealing (SA) for problem sizes ranging from 50 to 500 with $M = N$ and (a) c_i=rand[0.45,0.55], (b) c_i=0.5

NN, SA and LP approaches with the exact BB for an $N = M = 30$ problem. This is done both for non-homogeneous and homogeneous problems. The results are shown in table 5.2. As expected LP and in particular GH benefit from non-homogeneity both in respect of quality and of CPU (the latter is also very true for BB), while for homogeneous problems the NN algorithm is the winner.

For larger problems it is not feasible to use the exact BB algorithm. The best we can do is to compare the different approximate approaches, NN, SA and LP. The conclusions for problem sizes ranging from 50 to 500 are the same as above. The real strength of the NN approach is best exploited for the more homogeneous problems.

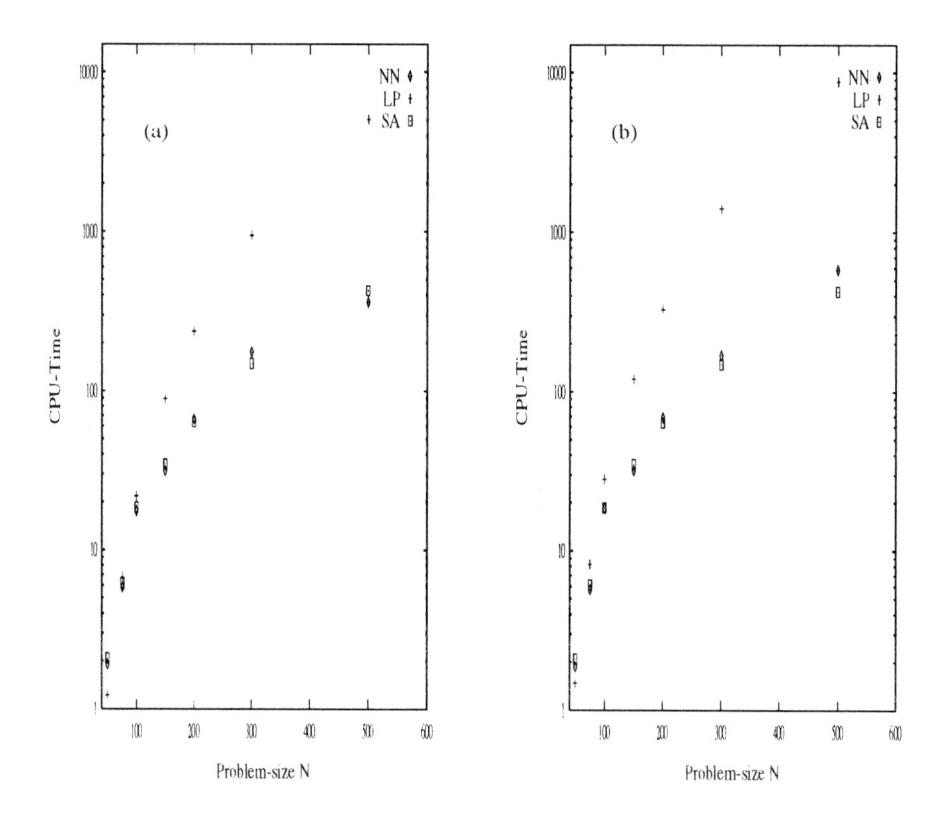

Figure 5.16: CPU consumption of the neural network (NN) and linear programming approaches (LP) normalized to simulated annealing (SA) for problem sizes ranging from 50 to 500 with $M = N$ and (a) c_i=rand[0.45,0.55], (b) c_i=0.5. The numbers refer to DEC3100 workstations.

Figures 5.15 and 5.16 show the performance and CPU consumption for $N \in [50, 500]$ with $M = N$. In summary the MFT approach is very competitive as compared to other approximate methods for the hard homogeneous problems, both with respect to solution quality and time consumption. It also compares very well with exact solutions for problem sizes where these are accessible.

Being able to find good approximate solutions to difficult knapsack problems opens up several application areas within the field of resource allocation. Hence the achievements noted in this section could be very important.

5.10 Summary

The neural approach is in its nature very different from other approximate methods. It has its roots in a statistical mechanics treatment of systems with many degrees of freedom, as has simulated annealing in general. With the MFT approximation it feels its way in a 'fuzzy' manner towards good solutions. This is in contrast to most other methods, where discrete moves take place in a solution space. The basic steps for solving combinatorial optimization problems with an ANN are:

- Map the problem onto a neural network by a suitable coding of the possible solutions and an appropriate choice of energy function. Where applicable, use Potts encoding rather than Ising neuron multiplexing.

- Make as much use as possible of prior knowledge about phase transition properties from analyzing the linearized dynamics. Special care is needed when defining the MFT equations in cases when the penalty terms are non-polynomial, as in the case of inequality constraints.

- Solve the corresponding mean field equations iteratively, under annealing in T.

- When the MFT equations have converged the solutions are checked with respect to 'legality'—whether or not they satisfy the basic constraints. If not, one either supplements the algorithm with a greedy heuristic, or re-anneals the system (perhaps with modified constraint coefficients).

For low-dimensional geometrical problems like the travelling salesman problem, it is often advantageous to use an alternative procedure, the deformable templates method. This reduces the number of degrees of freedom and hence the computing time.

With respect to quality the ANN method in general gives results comparable to those of other state-of-the-art approximative approaches, such as simulated annealing. In full-sized high school scheduling applications comparisons have not been possible since no results exist from other approaches.

The neural language is very natural for encoding many optimization problems like scheduling. The approach has the advantage of being easily used for revision of solutions due to new situations. One simply re-anneals the network with non-revisable units clamped.

As with all other ANN applications the neural optimization approach has the advantage of being fully parallelizable.

References

[1] E.R.Kandel and J.H.Schwartz (1991) *Principles of Neural Science*, 3rd ed. Elsevier Publishers, New York.

[2] J.Hertz, A.Krogh and R.G.Palmer (1991) *Introduction to the Theory of Neural Computation.* Addison-Wesley, Redwood City, Ca.

[3] D.E.Rumelhart and J.L.McClelland (Eds.) (1986) *Parallel Distributed Processing: Explorations in the Microstructure of Cognition (Vol. 1)*, MIT Press.

[4] A.Lapedes and R.Farber (1987) *Nonlinear signal processing using neural networks: prediction and system modeling.* Los Alamos Report LA-UR 87-2662.

[5] A.Weigend, B.A.Huberman and D.Rumelhart (1990) Predicting the future: a connectionist approach. *International J. of Neural Systems*, **3**, 193.

[6] J.J.Hopfield (1982) Neural networks and physical systems with emergent collective computational abilities. *Proc. of the National Academy of Science, USA*, **79**, 2554.

[7] J.J.Hopfield and D.W.Tank (1985) Neural computation of decisions in optimization problems. *Biol. Cyber.*, **52**, 141.

[8] C.Peterson and B.Söderberg (1989) A new method for mapping optimization problems onto neural networks. *International J. of Neural Systems*, **1**, 3.

[9] R.Durbin and D.Willshaw (1987) An analog approach to the travelling salesman problem using an elastic net method. *Nature*, **326**, 689.

[10] S.Kirkpatrick, C.D.Gelatt and M.P.Vecchi (1983) Optimization by simulated annealing. *Science*, **220**, 671.

[11] D.Chandler (1987) *Introduction to Modern Statistical Mechanics.* Oxford University Press, Oxford.

[12] C.Peterson and J.R.Anderson (1988) Neural networks and NP-complete optimization problems: a performance study on the graph bisection problem. *Complex Systems*, **2**, 59.

[13] C.Peterson (1990). Parallel distributed approaches to combinatorial optimization problems—benchmark studies on TSP. *Neural Computation*, **2**, 261.

[14] M.Y.Choi and B.A.Huberman (1983) Digital dynamics and the simulation of magnetic systems. *Physical Review*, **B28**, 2547.

[15] F.Y.Wu (1983) The Potts model. *Review of Modern Physics*, **54**, 235.

[16] L.Gislén, B.Söderberg and C.Peterson (1989) Teachers and classes with neural networks. *International J. of Neural Systems*, **1**, 167.

[17] L.Gislén, B.Söderberg and C.Peterson (1991) Scheduling high schools with neural networks. *Neural Computation*, (submitted).

[18] A.L.Yuille (1990) Generalized deformable models, statistical physics, and matching problems. *Neural Computation*, **2**, 1.

[19] A.Yuille, K.Honda and C.Peterson (1991) Particle tracking by deformable templates. *Proc. IEEE INNS International Joint Conference on Neural Networks*, Seattle, WA.

[20] T.Kohonen (1982) Self organized formation of topologically correct feature maps. *Biol.Cyber.*, **43**, 59.

[21] T.Kohonen (1990) *Self-organization and Associative Memory*, 3rd ed. Springer-Verlag, Berlin, Heidelberg.

[22] M.Ohlsson, C.Peterson and A.Yuille (1992) Track finding with deformable templates—the elastic arms approach. *Computer Physics Comm.*, (to appear).

[23] M.Gyulassy and H.Harlander (1991) Elastic tracking and neural network algorithms for complex pattern recognition. *Computer Physics Comm.*, **66**, 31.

[24] M.Ohlsson, C.Peterson and B.Söderberg (1991) Neural networks for optimization problems with inequality constraints—the knapsack problem. *Neural Computation*, (submitted).

[25] W.P.Press, B.P.Flannery, S.A.Teukolsky and W.T.Vettering (1986) *Numerical Recipes, The Art of Scientific Computing.* Cambridge University Press, Cambridge.

Chapter 6

Lagrangean Relaxation

John E Beasley

6.1 Introduction

As was remarked in chapter 1 of this book, finding good solutions to hard problems in combinatorial optimization requires the consideration of two issues:

- calculation of an upper bound that is as close as possible to the optimum;
- calculation of a lower bound that is as close as possible to the optimum.

General techniques for generating good upper bounds are essentially heuristic methods of the type considered in the preceding chapters of this book. In addition, for any particular problem, we may well have techniques which are specific to the problem being solved.

On the question of lower bounds, one well-known general technique which is available is Linear Programming (LP) relaxation. In LP relaxation we take an integer (or mixed-integer) programming formulation of the problem and relax the integrality requirement on the variables. This gives a linear program which can either be:

- solved exactly using a standard algorithm (simplex or interior point); or
- solved heuristically (dual ascent).

The solution value obtained for this linear program gives a lower bound on the optimal solution to the original problem. We shall illustrate both of these approaches in this chapter.

Another well-known (and well-used) technique which is available to find lower bounds is Lagrangean relaxation. This technique will be expounded at much greater length. Suffice to say for the moment that Lagrangean relaxation involves:

1. taking an integer (or mixed-integer) programming formulation of the problem;

2. attaching Lagrange multipliers to some of the constraints in this formulation and relaxing these constraints into the objective function;

3. solving (exactly) the resulting integer (or mixed-integer) program.

The solution value obtained from step 3 above gives a lower bound on the optimal solution to the original problem. At first sight this might not appear to be a useful approach since at step 1 above we have an integer (or mixed-integer) programming formulation of the problem and we propose to generate a lower bound for it by solving another integer (or mixed-integer) program (step 3).

There are two basic reasons why this approach is well-known (and well-used).

- Many combinatorial optimization problems consist of an easy problem (in the NP-complete sense, i.e. solvable by a polynomially bounded algorithm) complicated by the addition of extra constraints. By absorbing these complicating constraints into the objective function (step 2) we are left with an easy problem to solve and attention can then be turned to choosing numerical values for the Lagrange multipliers.

- Practical experience with Lagrangean relaxation has indicated that it gives very good lower bounds at reasonable computational cost.

Choosing values for the Lagrange multipliers is of key importance in terms of the quality of the lower bound generated (we much prefer lower bounds which are close to the optimal solution). Two general techniques are available here—*subgradient optimization* and *multiplier adjustment*.

Readers are reminded here of the description and diagrams (Figures 1.1, 1.2) given in chapter 1. For finding upper bounds we have heuristic techniques such as those described in chapters 2-5. For finding lower bounds we have the various relaxations mentioned above.

6.2 Overview

6.2.1 Techniques

In this chapter we deal with:

- Lagrangean relaxation
- Lagrangean heuristics
- problem reduction
- subgradient optimization
- multiplier adjustment
- dual ascent
- tree search

These techniques are building blocks for constructing optimal solution (exact) algorithms, that is algorithms which guarantee to find the optimal solution to a particular problem.

But guaranteeing to find the optimal solution is not sufficient. An algorithm that guarantees to find the optimal solution, but which takes days to run on a computer, is obviously of much less use than an algorithm which guarantees to find the optimal solution but which takes only minutes to run on a computer.

This issue of computational effectiveness, i.e. building an algorithm that can solve problems in a reasonable time frame, is a key one in algorithm design and development. Therefore, in this chapter, we also have as an implicit objective the design of a computationally effective optimal solution algorithm incorporating the above techniques. In order to do this we shall consider one of the simplest

NP-complete combinatorial optimization problems—the set covering problem.

6.2.2 Review of the literature

Much of what is contained in this chapter is historical—that is, it has been known for a number of years. To enable the reader to get some insight into current up-to-date work involving Lagrangean relaxation we have adopted the strategy of:

- selecting calendar year 1991 (the most recent complete calendar year at the time of writing);
- selecting a number of journals; and
- reviewing all papers in those journals in our selected year which use Lagrangean relaxation.

This will provide a systematic insight into the range of problems to which Lagrangean relaxation is being applied.

6.3 Basic Methodology

6.3.1 Introduction

Consider the following general zero-one problem (written in matrix notation) which we shall refer to as problem (P):

$$\begin{array}{ll} \text{minimize} & cx \\ \text{subject to} & Ax \geq b \\ & Bx \geq d \\ & x \in (0,1) \end{array}$$

Note here that although we deal in this chapter purely with zero-one integer programs the material presented is equally applicable both to pure (general) integer programs and to mixed-integer programs.

As mentioned above, one way to generate a lower bound on the optimal solution to problem (P) is via the linear programming relaxation. This entails replacing the integrality constraint [$x \in (0,1)$] by its linear relaxation [$0 \leq x \leq 1$] to give the following linear program:

$$\begin{array}{ll} \text{minimize} & cx \\ \text{subject to} & Ax \geq b \\ & Bx \geq d \\ & 0 \leq x \leq 1 \end{array}$$

This linear program can be solved exactly using a standard algorithm (e.g. simplex or interior point) and the solution value obtained gives a lower bound on the optimal solution to the original problem (P).

In many cases however solving the linear programming relaxation of (P) is impracticable, because typically (P) involves a large (often extremely large) number of variables and/or constraints. We therefore need alternative techniques for generating lower bounds.

6.3.2 Lagrangean relaxation

Lagrangean relaxation was developed in the early 1970s with the pioneering work of Held and Karp [1, 2] on the travelling salesman problem and is today an indispensable technique for generating lower bounds for use in algorithms to solve combinatorial optimization problems.

We define the Lagrangean relaxation of problem (P) with respect to the constraint set $Ax \geq b$ by introducing a Lagrange multiplier vector $\lambda \geq 0$ which is attached to this constraint set and brought into the objective function to give:

$$\begin{array}{ll} \text{minimize} & cx + \lambda(b - Ax) \\ \text{subject to} & Bx \geq d \\ & x \in (0, 1) \end{array}$$

What we have done here is:

- to choose some set of constraints in the problem for relaxation; and
- to attach Lagrange multipliers to these constraints in order to bring them into the objective function.

The key point is that the program we are left with after Lagrangean relaxation, for any $\lambda \geq 0$, gives a lower bound on the optimal solution to the original problem (P). This can be seen as follows.

The value of

$$\begin{array}{ll} \text{minimize} & cx \\ \text{subject to} & Ax \geq b \\ & Bx \geq d \\ & x \in (0,1) \end{array}$$

is greater than the value of

$$\begin{array}{ll} \text{minimize} & cx + \lambda(b - Ax) \\ \text{subject to} & Ax \geq b \\ & Bx \geq d \\ & x \in (0,1) \end{array}$$

since as $\lambda \geq 0$ and $(b - Ax) \leq 0$ we are merely adding a term which is ≤ 0 to the objective function. This in turn is greater than the value of

$$\begin{array}{ll} \text{minimize} & cx + \lambda(b - Ax) \\ \text{subject to} & Bx \geq d \\ & x \in (0,1) \end{array}$$

since removing a set of constraints from a minimization problem can only reduce the objective function value.

The program after Lagrangean relaxation, namely:

$$\begin{array}{ll} \text{minimize} & cx + \lambda(b - Ax) = (c - \lambda A)x + \lambda b \\ \text{subject to} & Bx \geq d \\ & x \in (0,1) \end{array}$$

can be called the Lagrangean lower bound program (LLBP) since, as shown above, it provides a lower bound on the optimal solution to the original problem (P) for any $\lambda \geq 0$. Note that the above proof that Lagrangean relaxation generates lower bounds is quite general, i.e. the constraints/objective function need not be linear functions.

There are two key issues highlighted by the above Lagrangean relaxation:

- a *strategic* issue—namely, why did we choose to relax the set of constraints $Ax \geq b$ when we could equally well have chosen to relax $Bx \geq d$?

- a *tactical* issue—namely, how can we find numerical values for the multipliers?

  In particular note here that we are interested in finding the values for the multipliers that give the maximum lower bound, i.e. the lower bound that is as close as possible to the value of the optimal integer solution. This involves finding multipliers which correspond to:

$$\max_{\lambda \geq 0} \left\{ \begin{array}{ll} \text{minimize} & cx + \lambda(b - Ax) \\ \text{subject to} & Bx \geq d \\ & x \in (0,1) \end{array} \right\}$$

  This program is called the *Lagrangean dual program.*

  Ideally the optimal value of the Lagrangean dual program (a maximization program) is equal to the optimal value of the original zero-one integer program (a minimization problem). If the two programs do not have optimal values which are equal then a *duality gap* is said to exist, the size of which is measured by the (relative) difference between the two optimal values.

In order to illustrate Lagrangean relaxation we shall now consider one of the simplest NP-complete combinatorial optimization problems—the set covering problem.

6.3.3 Set covering problem

The set covering problem (SCP) is the problem of covering the rows of an m-row, n-column, zero-one matrix $[a_{ij}]$ by a subset of the columns at minimum cost.

Defining $x_j = 1$ if column j (cost $c_j > 0$) is in the solution
$\qquad = 0$ otherwise

the SCP is:

$$\begin{array}{ll} \text{minimize} & \sum_{j=1}^{n} c_j x_j \\ \text{subject to} & \sum_{j=1}^{n} a_{ij} x_j \geq 1; \;\; i = 1, \ldots, m \\ & x_j \in (0,1); \;\; j = 1, \ldots, n \end{array}$$

The first set of constraints in this program ensures that each row is covered by at least one column and the second set form the integrality constraints.

In order to generate a Lagrangean relaxation of this SCP we need:

1. to choose some set of constraints in the problem for relaxation; and

2. to attach Lagrange multipliers to these constraints in order to bring them into the objective function.

Step 1 above is not usually an easy step. As commented above, the choice of which set of constraints to relax is a strategic issue. However, for the SCP we simply have one distinct set of constraints $(\sum_{j=1}^{n} a_{ij}x_j \geq 1; i = 1, \ldots, m)$ and so we choose this set of constraints for relaxation and attach Lagrange multipliers $\lambda_i \geq 0$ $(i = 1, \ldots, m)$ to these constraints.

If we do this we find that LLBP is:

$$\begin{array}{ll} \text{minimize} & \sum_{j=1}^{n} c_j x_j + \sum_{i=1}^{m} \lambda_i (1 - \sum_{j=1}^{n} a_{ij} x_j) \\ \text{subject to} & x_j \in (0,1); \;\; j = 1, \ldots, n \end{array}$$

i.e.

$$\begin{array}{ll} \text{minimize} & \sum_{j=1}^{n} [c_j - \sum_{i=1}^{m} \lambda_i a_{ij}] x_j + \sum_{i=1}^{m} \lambda_i \\ \text{subject to} & x_j \in (0,1); \;\; j = 1, \ldots, n \end{array}$$

Defining $C_j = [c_j - \sum_{i=1}^{m} \lambda_i a_{ij}]$, $j = 1, \ldots, n$, we find that LLBP becomes:

$$\begin{array}{ll} \text{minimize} & \sum_{j=1}^{n} C_j x_j + \sum_{i=1}^{m} \lambda_i \\ \text{subject to} & x_j \in (0,1); \;\; j = 1, \ldots, n \end{array}$$

Now the solution (X_j) to LLBP can be found by inspection, namely:

$$\begin{aligned} X_j &= 1 \quad \text{if } C_j \leq 0 \\ &= 0 \quad \text{otherwise} \end{aligned}$$

with the solution value (Z_{LB}) of LLBP being given by:

$$Z_{LB} = \sum_{j=1}^{n} C_j X_j + \sum_{i=1}^{m} \lambda_i$$

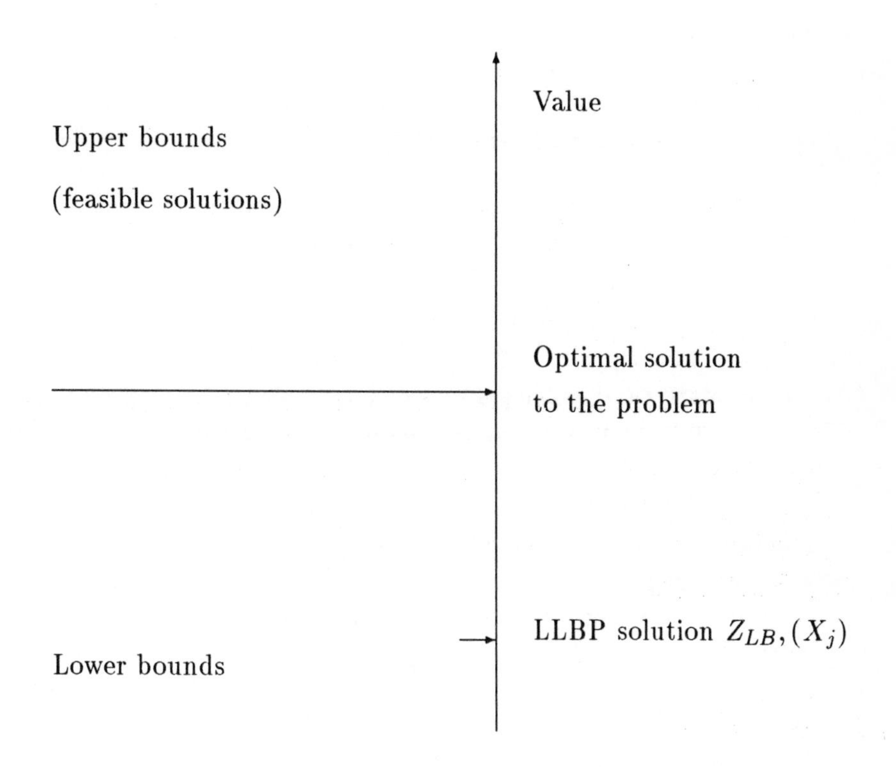

Figure 6.1: LLBP solution

where Z_{LB} is a lower bound on the optimal solution to the original SCP.

Figure 6.1 summarizes the situation. In that figure we have a point on the value line (a lower bound) associated with the solution $(Z_{LB},(X_j))$ to LLBP.

To illustrate the Lagrangean relaxation of the SCP given above consider the following example SCP (with 3 rows and 4 columns) defined by:

$$(c_j) = (2,3,4,5)$$

$$[a_{ij}] = \begin{bmatrix} 1010 \\ 1001 \\ 0111 \end{bmatrix}$$

i.e. the example SCP is:

$$\begin{array}{ll} \text{minimize} & 2x_1 + 3x_2 + 4x_3 + 5x_4 \\ \text{subject to} & x_1 + x_3 \geq 1 \\ & x_1 + x_4 \geq 1 \\ & x_2 + x_3 + x_4 \geq 1 \\ & x_j \in (0,1); \;\; j = 1, \ldots, 4 \end{array}$$

Note here that the optimal solution to this SCP is of value 5 with $x_1 = x_2 = 1$ and $x_3 = x_4 = 0$.

To generate the Lagrangean lower bound program we attach Lagrange multipliers $\lambda_i \geq 0; i = 1,2,3$ to the three constraints in this SCP to get:

$$\begin{array}{lll} \text{minimize} & 2x_1 + 3x_2 + 4x_3 + 5x_4 & +\lambda_1(1 - x_1 - x_3) \\ & & +\lambda_2(1 - x_1 - x_4) \\ & & +\lambda_3(1 - x_2 - x_3 - x_4) \\ \text{subject to} & x_j \in (0,1); \;\; j = 1, \ldots, 4 & \end{array}$$

i.e. LLBP is

$$\begin{array}{ll} \text{minimize} & C_1x_1 + C_2x_2 + C_3x_3 + C_4x_4 + \lambda_1 + \lambda_2 + \lambda_3 \\ \text{subject to} & x_j \in (0,1); \;\; j = 1, \ldots, 4 \end{array}$$

where

$$\begin{aligned} C_1 &= (2 - \lambda_1 - \lambda_2) \\ C_2 &= (3 - \lambda_3) \\ C_3 &= (4 - \lambda_1 - \lambda_3) \\ C_4 &= (5 - \lambda_2 - \lambda_3) \end{aligned}$$

Hence (X_j), the solution values of the (x_j), are given by

$$\begin{aligned} X_j &= 1 \quad \text{if } C_j \leq 0 \\ &= 0 \quad \text{otherwise} \end{aligned}$$

with the solution value for LLBP (Z_{LB})—a valid lower bound on the optimal solution to the original SCP—being given by

$$Z_{LB} = C_1X_1 + C_2X_2 + C_3X_3 + C_4X_4 + \lambda_1 + \lambda_2 + \lambda_3$$

6.3.4 Example Lagrange multiplier values

As commented above the choice of numerical values for the Lagrange multipliers is a tactical issue. For the moment consider the (arbitrarily decided) set of values for the Lagrange multipliers of:

$$\lambda_1 = 1.5$$

$$\lambda_2 = 1.6$$

$$\lambda_3 = 2.2$$

then

$$C_1 = (2 - \lambda_1 - \lambda_2) = -1.1$$

$$C_2 = (3 - \lambda_3) = 0.8$$

$$C_3 = (4 - \lambda_1 - \lambda_3) = 0.3$$

$$C_4 = (5 - \lambda_2 - \lambda_3) = 1.2$$

The solution to LLBP is

$$X_1 = 1, X_2 = X_3 = X_4 = 0$$

from which it is easily seen that $Z_{LB} = -1.1+0+0+0+1.5+1.6+2.2 = 4.2$, which is indeed a lower bound on the optimal solution (which we know is of value 5) to the original SCP.

6.3.5 Advanced Lagrangean relaxation

The basic Lagrangean relaxation methodology can be extended in several ways. In this section we draw some of these results together, and comment on their implications.

(1) If we relax equality constraints then λ_i is unrestricted in sign (i.e. λ_i can be positive or negative).

(2) A common fallacy in Lagrangean relaxation is to believe that, if the solution to LLBP is feasible for the original problem, then it is also optimal for the original problem. This is incorrect.

For example consider the SCP with 3 rows and 4 columns dealt with above. Set $\lambda_1 = \lambda_2 = \lambda_3 = 10$ and solve LLBP. The solution is $X_1 = X_2 = X_3 = X_4 = 1$. This is certainly a feasible solution for the original problem (the SCP) but by no means the optimal solution!

Under what circumstances therefore does the solution to LLBP being feasible for the original problem also imply that it is optimal for the original problem?

The answer to this question is simple. Consider LLBP:

$$\begin{array}{ll} \text{minimize} & cx + \lambda(b - Ax) \\ \text{subject to} & Bx \geq d \\ & x \in (0,1) \end{array}$$

Suppose that the Lagrange multipliers $\lambda \geq 0$ are such that the solution X to LLBP is feasible for the original problem (i.e. X satisfies $AX \geq b$, $BX \geq d$ and $X \in (0,1)$). The value of this feasible solution is cX whereas the value of the lower bound obtained from LLBP is $[cX + \lambda(b - AX)]$. Then if these two values coincide, i.e. the upper bound cX is equal to the lower bound $[cX + \lambda(b - AX)]$, X is optimal.

In other words a solution X to a Lagrangean lower bound program is optimal for the original problem *only* if:

1. X is feasible for the original problem; and
2. $cX = [cX + \lambda(b - AX)]$, i.e. $\lambda(b - AX) = 0$

The reason why the fallacy referred to above has appeared is clear. If we are relaxing *equality* constraints ($Ax = b$) then any solution to the Lagrangean lower bound program which is feasible for the original problem automatically satisfies both 1 and 2 above and so is optimal.

(3) If the solution to LLBP (for all possible multiplier (λ) values) is unchanged by replacing the integrality constraint $[x \in (0,1)]$ in LLBP by its linear relaxation $[0 \leq x \leq 1]$ then the Lagrangean relaxation/Lagrangean lower bound program is said to have the *integrality property*.

To illustrate this consider the Lagrangean relaxation of the SCP given above, which was:

$$\begin{array}{ll} \text{minimize} & \sum_{j=1}^{n} C_j x_j + \sum_{i=1}^{m} \lambda_i \\ \text{subject to} & x_j \in (0,1); \; j = 1,\ldots,n \end{array}$$

with solution

$$\begin{aligned} X_j &= 1 \quad \text{if } C_j \leq 0 \\ &= 0 \quad \text{otherwise} \end{aligned}$$

It is clear that replacing $x_j \in (0,1); j = 1,\ldots,n$ by $0 \leq x_j \leq 1; j = 1,\ldots,n$ leaves the solution unchanged.

Hence the Lagrangean relaxation of the SCP given above does have the integrality property.

(4) If the Lagrangean relaxation has the integrality property then the maximum lower bound attainable from LLBP is *equal* to the value of the linear programming relaxation of the original problem.

Hence for the Lagrangean relaxation of the SCP considered above, the maximum lower bound attainable from LLBP (i.e. the value of the Lagrangean dual program) is equal to the value of the LP relaxation of the original problem.

(5) If the Lagrangean relaxation does not have the integrality property then the maximum lower bound attainable from LLBP is *greater than (or equal to)* the value of the linear programming relaxation of the original problem.

(6) If a problem has a number of different sets of constraints then there are a number of possible Lagrangean relaxations. This is the strategic choice issue mentioned above, namely:

Which set (or sets) of constraints should we choose to relax?

The integrality property defined above provides a key mechanism to assist us in choosing the Lagrangean relaxation to use. It gives an indication of how good the maximum (best) lower bound we can get from a particular relaxation is.

Suppose we ask the question 'Does a Lagrangean relaxation have the integrality property?'. Then we have:

Answer: yes $\Rightarrow$ maximum lower bound $=$ LP relaxation solution
no $\Rightarrow$ maximum lower bound $\geq$ LP relaxation solution

So a Lagrangean relaxation without the integrality property is to be preferred as it offers the prospect of getting better lower bounds.

But lower bounds from Lagrangean relaxations are not free. Computational effort is needed to solve LLBP for each set of values for the multipliers and, as we shall see below, we may have to solve LLBP for many different sets of multipliers.

Essentially in choosing the Lagrangean relaxation to use we have to balance the computational effort involved in solving LLBP against the maximum lower bound theoretically attainable (note that we may not, computationally, actually attain this bound).

A structured way to approach this strategic choice issue of which Lagrangean relaxation to use is by means of a simple table.

Consider the original problem (P):

$$\begin{array}{ll} \text{minimize} & cx \\ \text{subject to} & Ax \geq b \\ & Bx \geq d \\ & x \in (0,1) \end{array}$$

then we can set up the table:

Relaxation number	Constraints		Computational effort involved in solving LLBP		Integrality property?
	$Ax \geq b$ Relax?	$Bx \geq d$ Relax?	Number of multipliers	Number of operations	
1	no	no	?	?	?
2	no	yes	?	?	?
3	yes	no	?	?	?
4	yes	yes	?	?	?

For each set of constraints we have a choice of whether to relax them or not. Hence if there are s sets of constraints there are 2^s possible relaxations (in problem (P) above for example we have 2 sets of constraints so $s = 2$ and there are $2^2 = 4$ possible relaxations).

Filling in the details for 'Computational effort involved in solving LLBP' and 'Integrality property?' (the question marks in the above table) depends upon examining each relaxation in mathematical detail. Specifically we need to find:

- the number of Lagrange multipliers needed—this affects the amount of computer memory needed;

- the number of arithmetic operations required to solve LLBP (e.g. $O(n^2)$)—this affects the amount of computer time needed;
- whether LLBP has the integrality property or not.

Whilst the above table might seem somewhat daunting (e.g. 4 sets of constraints lead to $2^4 = 16$ possible relaxations), in practice some relaxations are not worth considering:

1. relaxation 1 in the table can be ignored since it is just the original problem (no constraints relaxed);
2. we need only consider relaxations for which 'Computational effort involved in solving LLBP' involves a polynomial (or pseudo-polynomial) number of operations, i.e. relaxations which are themselves 'hard' problems (in the NP-complete sense) are of no (computational) advantage;
3. all relaxations which have the integrality property have the same maximum lower bound, so the particular relaxation (with the integrality property) which involves the least computational effort in solving LLBP would seem the best choice to make. In particular note here that the relaxation which involves relaxing all constraints has the integrality property.

After 1, 2 and 3 above we are left with (hopefully) a small number of relaxations which are deserving of further investigation (either theoretically or computationally).

One hint that may be useful to researchers is to include a column in the above table relating to previous work dealing with each relaxation. This is helpful both in seeing where previous work has been concentrated and in seeing where there might be a legitimate 'gap' in previous research work.

(7) In fact there are many more Lagrangean relaxations of a problem than might be apparent from the table above. To see this consider the original problem (P):

$$\begin{array}{ll} \text{minimize} & cx \\ \text{subject to} & Ax \geq b \\ & Bx \geq d \\ & x \in (0,1) \end{array}$$

and transform this into:

$$\begin{array}{ll} \text{minimize} & cx \\ \text{subject to} & Ax \geq b \\ & y = x \\ & By \geq d \\ & x \in (0,1) \\ & y \in (0,1) \end{array}$$

Here we introduce an artificial zero-one variable y equal to x and replace x in one set of constraints by y. It is clear that the original problem (P) and this transformed problem are equivalent.

If we now relax the equality constraint linking y and x together by introducing a Lagrange multiplier vector λ (now unrestricted in sign) we get the Lagrangean lower bound program (LLBP):

$$\begin{array}{ll} \text{minimize} & cx + \lambda(x - y) \\ \text{subject to} & Ax \geq b \\ & By \geq d \\ & x \in (0,1) \\ & y \in (0,1) \end{array}$$

and it is clear that LLBP is *separable* into the sum of two programs:

$$\begin{array}{ll} \text{minimize} & (c + \lambda)x \\ \text{subject to} & Ax \geq b \\ & x \in (0,1) \end{array} \quad \textit{and} \quad \begin{array}{ll} \text{minimize} & -\lambda y \\ \text{subject to} & By \geq d \\ & y \in (0,1) \end{array}$$

The sum of the solutions to these two programs (for any value of λ) provides a lower bound on the optimal solution to the original (integer) problem.

This technique of introducing a 'copy' of the variables and then relaxing the equality constraint linking variables is known as *Lagrangean decomposition.* Informally it is clear that we can examine Lagrangean decomposition systematically by:

- introducing different artificial variables for all but one constraint (that constraint being expressed in terms of the original variables);

- adding all possible equality linking constraints between these artificial variables and the original variables (e.g. if we have three constraints involving original variables x then we have two artificial variables (y and z, say) and the set of possible equality linking constraints is $\{x = y, x = z, y = z\}$);

- considering the relaxation of all equality linking constraints in a Lagrangean fashion using a table such as shown above.

More formally if there are s sets of constraints then there are $s(s-1)/2$ equality linking constraints and hence $2^{s(s-1)/2}$ possible Lagrangean decompositions.

(8) It may be possible to add to LLBP constraints which would have been redundant (unnecessary) in the original integer (or mixed-integer) formulation of the problem, but which strengthen the lower bound obtained from LLBP (for any value of λ). For example, for the SCP:

$$\begin{array}{ll} \text{minimize} & \sum_{j=1}^{n} c_j x_j \\ \text{subject to} & \sum_{j=1}^{n} a_{ij} x_j \geq 1; \;\; i = 1, \ldots, m \\ & x_j \in (0,1); \;\; j = 1, \ldots, n \end{array}$$

since $c_j > 0$ $(j = 1, \ldots, n)$ it is clear that at least one, but no more than m, variables can take the value 1 in the optimal solution and so the constraint

$$1 \leq \sum_{j=1}^{n} x_j \leq m$$

is valid (but redundant in the original formulation).

Whereas the LLBP for the SCP was given before by:

$$\begin{array}{ll} \text{minimize} & \sum_{j=1}^{n} C_j x_j + \sum_{i=1}^{m} \lambda_i \\ \text{subject to} & x_j \in (0,1); \;\; j = 1, \ldots, n \end{array}$$

adding this constraint to LLBP gives:

$$\begin{array}{ll} \text{minimize} & \sum_{j=1}^{n} C_j x_j + \sum_{i=1}^{m} \lambda_i \\ \text{subject to} & 1 \leq \sum_{j=1}^{n} x_j \leq m \\ & x_j \in (0,1); \;\; j = 1, \ldots, n \end{array}$$

It is clear that:

- the lower bound obtained from this new (strengthened) LLBP is always (for any set of (λ_i)) better $(\geq)$ than the lower bound obtained from the original LLBP; and

- this new (strengthened) LLBP is computationally easy to solve: sort the variables x_j by increasing C_j order and set $x_j = 1$ for at least one, and at most m, variables in this list depending upon whether $C_j \leq 0$ or not.

6.4 Lagrangean Heuristics and Problem Reduction

The solution to LLBP can be used to construct feasible solutions to the original problem (a Lagrangean heuristic) and also to reduce the size of the problem that we need to consider (problem reduction). We deal with these two topics below.

6.4.1 Lagrangean heuristic

In a Lagrangean heuristic we take the solution to LLBP and attempt to convert it into a feasible solution for the original problem by suitable adjustment (if necessary). This feasible solution constitutes an upper bound on the optimal solution to the problem (cf. Figure 6.1). To illustrate the concept of a Lagrangean heuristic we will develop a Lagrangean heuristic for the SCP.

In the set covering problem all feasible solutions consist of a set of columns (x_j) which cover each row at least once. In the solution to LLBP for the SCP we have some X_j one and some X_j zero. This may result in some rows not being covered; plainly these rows need to be covered to constitute a feasible solution for the SCP.

Hence one possible (very simple) Lagrangean heuristic is to construct a feasible solution S to the original SCP in the following way:

1. set $S = \{j | X_j = 1; j = 1, \ldots, n\}$
2. for each row i which is uncovered (i.e. $\sum_{j \in S} a_{ij} X_j = 0$) add the column corresponding to $\min_{j=1,\ldots,n}\{c_j | a_{ij} = 1\}$ to S
3. S will now be a feasible solution to the original SCP of cost $\sum_{j \in S} c_j$.

Note that the key feature of a Lagrangean heuristic is that we are building upon the current solution to LLBP. The essential idea here is that just as the solution value for LLBP gives us useful information (a lower bound on the optimal solution to the original problem) so the *structure* of the solution to LLBP (i.e. the value of the variables) may well be giving us useful information about the structure of the optimal solution.

To illustrate the Lagrangean heuristic given above consider the example LLBP solution of $X_1 = 1, X_2 = X_3 = X_4 = 0$ that we had before. Applying this Lagrangean heuristic to our example LLBP solution we get:

1. $S = \{1\}$

2. row 3 is the only uncovered row and the minimum cost column covering this row is column 2 so add column 2 to S

3. $S = \{1, 2\}$ is now a feasible solution to the original SCP of cost $c_1 + c_2 = 2 + 3 = 5$.

Fortuitously here our Lagrangean heuristic has actually found the optimal solution to the original problem. Obviously this may not happen in all cases. However each time we solve LLBP the Lagrangean heuristic has an opportunity to transform the solution to LLBP into a feasible solution for the original problem. If, as is common in practice (see below), we solve LLBP many times then the Lagrangean heuristic has many opportunities to transform the solution to LLBP into a feasible solution for the original problem.

Designing a Lagrangean heuristic for a particular LLBP is an art, the success of which is judged solely by computational performance, i.e. whether a particular Lagrangean heuristic gives good quality (near-optimal or optimal) solutions in a reasonable computation time.

Our experience, based upon applying Lagrangean heuristics to a number of different problems (e.g. see [3, 4, 5, 6, 7, 8, 9, 10]) has been that relatively simple Lagrangean heuristics can give good quality results.

6.4.2 Problem reduction

Problem reduction consists of fixing the values of certain variables in the problem, enabling a reduction in problem size to be achieved. For example for the SCP:

- fixing $x_k = 1$ for some k enables us to delete column k and all rows covered by column k from the problem (since once rows have been covered by some column they need no longer be considered);

- fixing $x_k = 0$ for some k enables us to delete column k from the problem.

Let: R_0 be the set of variables fixed to zero;
R_1 be the set of variables fixed to one.

The basic idea here is to make reductions such that combining the optimal solution to the reduced problem with:

$$x_k = 0; \quad \forall k \in R_0$$
$$x_k = 1; \quad \forall k \in R_1$$

gives a solution to the original problem which is *optimal* for the original problem: i.e. the optimal solution to the original problem can be found by combining the optimal solution to the reduced problem with the reduction information.

This is important because our aim is to develop an algorithm that guarantees to find the optimal solution to the (original) set covering problem. The essential reason for doing problem reduction is the *hope* that the reduced problem will be quicker to solve in terms of computation time.

In combinatorial optimization we are often dealing with large problems with an enormous number of potential solutions. For example the SCP has $O(2^n)$ potential solutions (some of which may be infeasible). We need to find the (possibly unique) optimal solution. To give an idea of the size of the task, a current state-of-the-art algorithm for the set covering problem has solved problems with n=3000. Readers can tax their own calculators in finding a value for 2^{3000}!

Informally, one can think of looking for the optimal solution to a combinatorial optimization problem as analogous to looking for a needle in a haystack. Our task is to search the haystack to find the needle. By making a reduction in problem size, thereby eliminating potential solutions, we are throwing away portions of the haystack, reducing the amount that we have to search, in the hope that this will mean finding the needle more quickly!

To see how problem reduction can be achieved let Z_{UB} be the value of the best feasible solution found (e.g. by a Lagrangean heuristic) for the SCP we are considering.

Fixing $x_k = 0$

Problem reduction of the first type, involving fixing $x_k = 0$, can be carried out as follows:

1. From the original SCP and LLBP it is clear that imposing the additional constraint that a particular column k must be in the optimal solution ($x_k = 1$) results in a lower bound of $Z_{LB} + C_k$ for each column k for which $X_k = 0$. (Hint: solve LLBP with the additional constraint that $x_k = 1$).

2. Thus we can remove column k from the problem if:

   (a) $X_k = 0$; and

   (b) $Z_{LB} + C_k > Z_{UB}$

   since a column k with $Z_{LB}+C_k$ (the lower bound corresponding to an optimal solution containing column k) greater than Z_{UB} cannot be in the optimal solution.

Figure 6.2 illustrates the situation. In that figure we have two points on the value line. One, an upper bound, Z_{UB} and the other, a lower bound, associated with the solution $(Z_{LB}, (X_j))$ to LLBP. Currently the (unknown) optimal solution to the problem lies somewhere in the interval $[Z_{LB}, Z_{UB}]$ (possibly at one of the two end-points of the interval).

Suppose that column k were to be part of the optimal solution (i.e. $x_k = 1$), then this solution would have to lie in the interval $[Z_{LB} + C_k, Z_{UB}]$. But we know that $Z_{LB} + C_k > Z_{UB}$ so we have a contradiction; thus column k cannot be part of the optimal solution. It can therefore be removed and the optimal solution to the reduced problem will be the same as the optimal solution to the original problem.

To illustrate problem reduction, consider our example LLBP solution with $X_1 = 1, X_2 = X_3 = X_4 = 0$ and $Z_{LB} = 4.2$. Currently no variables have been eliminated by problem reduction so $R_0 = \emptyset$ and

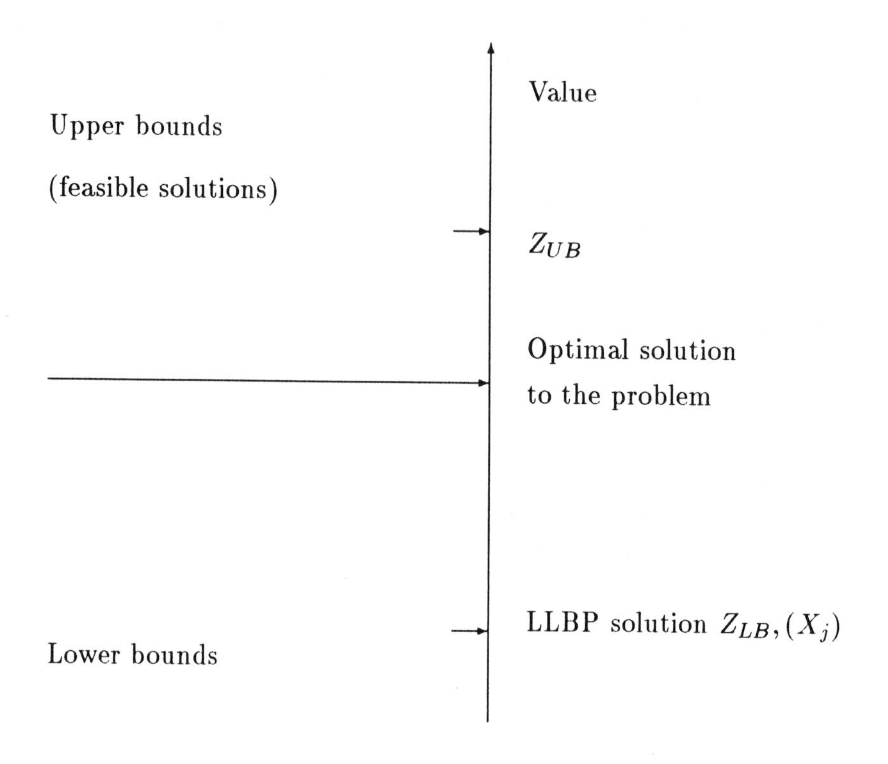

Figure 6.2: Upper and lower bounds

$R_1 = \emptyset$. For this solution we had $C_1 = -1.1, C_2 = 0.8, C_3 = 0.3$ and $C_4 = 1.2$.

We have found a feasible solution, via the Lagrangean heuristic, of value 5 so that $Z_{UB} = 5$. Hence:

$$\begin{aligned} Z_{LB} + C_k &= 4.2 + 0.8 = 5.0 \quad \text{for } k = 2 \text{ (as } X_2 = 0) \\ &= 4.2 + 0.3 = 4.5 \quad \text{for } k = 3 \text{ (as } X_3 = 0) \\ &= 4.2 + 1.2 = 5.4 \quad \text{for } k = 4 \text{ (as } X_4 = 0) \end{aligned}$$

Therefore column 4 cannot be in the optimal solution as $X_4 = 0$ and $Z_{LB} + C_4 > Z_{UB}$. Hence $R_0 = R_0 \cup [x_4]$ and column 4 can be deleted from the problem.

Fixing $x_k = 1$

The second type of problem reduction, fixing $x_k = 1$, can be carried out as follows:

1. From the original SCP and LLBP, it is clear that imposing the additional constraint that a particular column k cannot be in the optimal solution ($x_k = 0$) results in a lower bound of $Z_{LB} - C_k$ if $X_k = 1$. (Hint: solve LLBP with the additional constraint that $x_k = 0$).

2. We can therefore fix column k in the solution (fix $x_k = 1$) if:

   (a) $X_k = 1$; and

   (b) $Z_{LB} - C_k > Z_{UB}$

   since a column k with $Z_{LB} - C_k$ (the lower bound corresponding to an optimal solution without column k) greater than Z_{UB} cannot be in the optimal solution.

The proof of the correctness of the above argument follows exactly the same lines as that for the first type of problem reduction, and will be left to the reader.

To illustrate problem reduction fixing $x_k = 1$ consider our example LLBP solution as shown above. We find that

$$Z_{LB} - C_k = 4.2 - (-1.1) = 5.3 \text{ for } k = 1 \text{ (as } X_1 = 1)$$

Therefore column 1 must be in the optimal solution as $X_1 = 1$ and $Z_{LB} - C_1 > Z_{UB}$. Hence $R_1 = R_1 \cup [x_1]$, and column 1 and the rows that it covers (rows 1 and 2) can be deleted from the problem.

Our experience, based upon applying problem reduction to a number of different problems (e.g. see [3, 4, 5, 6, 7, 8, 9, 10]), has been that problem reduction significantly improves the computational performance of any algorithm. For example the algorithm given in [7] for the SCP includes problem reduction and typically over 75% (often well over 75%) of the variables are eliminated by problem reduction.

6.4.3 Remarks

In the preceding two subsections we have tried to illustrate how a Lagrangean lower bound program solution can be used to yield useful information about the optimal solution, both in terms of a feasible solution (Lagrangean heuristic), and in terms of problem reduction.

One encouraging point to note from the numerical examples given above is that even though the lower bound ($Z_{LB} = 4.2$) was 16% away from the optimal solution of 5 we were able to gain important information, namely:

- we actually found the optimal solution through the Lagrangean heuristic; and
- we made significant problem reduction (deleting column 4 from the problem and identifying column 1 as being in the optimal solution).

Whilst we will not always be so lucky, the point is clear—useful information can still be deduced from LLBP solutions which, in lower bound terms, are quite far from the optimal solution.

6.5 Determination of Lagrange Multipliers

In the previous section we have seen how to apply Lagrangean relaxation to:

- generate a lower bound;
- generate an upper bound (corresponding to a feasible solution); and
- reduce the size of the problem.

In addition we gave some guidelines upon the strategic issue of deciding which, out of all possible relaxations, to choose.

In this section we deal with the tactical issue, namely given a particular relaxation (i.e. the strategic choice has been made), how to find numerical values for the multipliers.

There are two basic approaches to deciding values for the Lagrange multipliers (λ_i)—subgradient optimization, and multiplier adjustment. We deal with each in turn.

6.5.1 Subgradient optimization

Recall the original problem that we are attempting to solve:

$$\begin{array}{ll} \text{minimize} & cx \\ \text{subject to} & Ax \geq b \\ & Bx \geq d \\ & x \in (0,1) \end{array}$$

The Lagrangean lower bound program (LLBP) for this problem was:

$$\begin{array}{ll} \text{minimize} & cx + \lambda(b - Ax) \\ \text{subject to} & Bx \geq d \\ & x \in (0,1) \end{array}$$

the solution to which, for any $\lambda \geq 0$, gives a lower bound on the optimal solution to the original (integer) problem.

Subgradient optimization is an iterative procedure which, from an initial set of Lagrange multipliers, generates further multipliers in a systematic fashion. It can be viewed as a procedure which attempts to maximize the lower bound value obtained from LLBP (i.e. to solve the Lagrangean dual program—see above) by suitable choice of multipliers.

Switching from matrix notation to summation notation, so that the relaxed constraints are

$$\sum_{j=1}^{n} a_{ij}x_j \geq b_i; \quad i = 1, \ldots, m$$

the basic procedure is as follows:

1. Let π be a user-defined parameter satisfying $0 < \pi \leq 2$. Initialize Z_{UB} (e.g. from some heuristic for the problem). Decide upon an initial set (λ_i) of multipliers.

2. Solve LLBP with the current set of multipliers (λ_i) , to get a solution (X_j) of value Z_{LB}.

3. Define *subgradients* G_i for the relaxed constraints, evaluated at the current solution, by:

$$G_i = b_i - \sum_{j=1}^{n} a_{ij}X_j; \quad i = 1, \ldots, m$$

4. Define a (scalar) step size T by

$$T = \frac{\pi(Z_{UB} - Z_{LB})}{\sum_{i=1}^{m} G_i^2}$$

This step size depends upon the gap between the current lower bound Z_{LB} and the upper bound Z_{UB} and the user defined parameter π (more of which below) with $\sum_{i=1}^{m} G_i^2$ being a scaling factor.

5. Update λ_i using

$$\lambda_i = \max(0, \lambda_i + TG_i); \quad i = 1, \ldots, m$$

and go to step 2 to resolve LLBP with this new set of multipliers.

As currently set out, this iterative procedure would never terminate, so we introduce a termination rule based either upon limiting the number of iterations that can be done, or upon the value of π where π is reduced during the course of the procedure. In the latter case, termination occurs when π is small—this will be discussed below.

We illustrate below one iteration of the subgradient optimization procedure for our example SCP.

1. Let $\pi = 2$.
   Let $Z_{UB} = 6$ (e.g. suppose we have applied some heuristic for the SCP and have found a feasible solution $x_1 = x_3 = 1, x_2 = x_4 = 0$, of value 6).
   Let $\lambda_1 = 1.5, \lambda_2 = 1.6, \lambda_3 = 2.2$, as before.

2. The solution to LLBP is $X_1 = 1, X_2 = X_3 = X_4 = 0$ with $Z_{LB} = 4.2$, as before.

3. The equations for the subgradients are:

$$\begin{aligned} G_1 &= (1 - X_1 - X_3) = 0 \\ G_2 &= (1 - X_1 - X_4) = 0 \\ G_3 &= (1 - X_2 - X_3 - X_4) = 1 \end{aligned}$$

4. The step size T is given by:

$$T = \frac{2(6 - 4.2)}{(0^2 + 0^2 + 1^2)} = 3.6$$

5. Updating λ_i gives:

$$\begin{aligned}
\lambda_1 &= \max(0, 1.5 + 3.6(0)) = 1.5 \\
\lambda_2 &= \max(0, 1.6 + 3.6(0)) = 1.6 \\
\lambda_3 &= \max(0, 2.2 + 3.6(1)) = 5.8
\end{aligned}$$

Resolving LLBP with this new set of multipliers gives $X_1 = X_2 = X_3 = X_4 = 1$ with a new lower bound of $Z_{LB} = -0.7$.

Note here that, in this case, changing the multipliers has made the lower bound worse than before (before it was 4.2, much closer to the optimal solution of 5 than the new value of -0.7). This behaviour is common in subgradient optimization; we cannot expect, and do not observe, a continual improvement in the lower bound at each iteration. Indeed, as seen above, the lower bound can even become negative.

However, suppose that we let Z_{max} be the *maximum* lower bound found over all subgradient iterations, where initially $Z_{max} = -\infty$, and we update Z_{max} at each subgradient iteration using $Z_{max} = \max(Z_{max}, Z_{LB})$. What has been observed computationally, by many workers in the field, is that Z_{max} increases quite rapidly during the initial subgradient iterations with the rate of increase slowing as many iterations are performed. In fact, it is common for Z_{max} to approach very close to (or even attain) the maximum lower bound possible from the LLBP, i.e. for Z_{max} to approach very close to (or even attain) the value of the Lagrangean dual program.

Figure 6.3 illustrates the situation as we perform subgradient iterations. As shown in that figure, we plot the lower bound found at each subgradient iteration on the value line. The best (maximum) of these lower bounds is Z_{max}. This is the lower bound closest to the optimal solution.

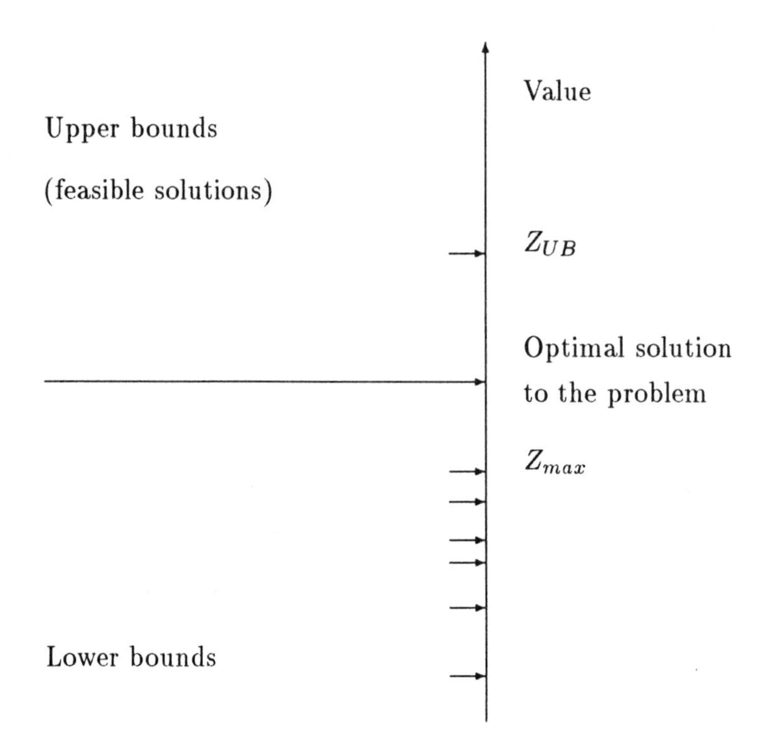

Figure 6.3: Subgradient optimization

6.5.2 Advanced subgradient optimization

Below we give a number of observations based upon our (extensive) experience with subgradient optimization.

(1) It is clear that the subgradient optimization procedure can be terminated if we find that $Z_{max} = Z_{UB}$. In this case the value of the maximum lower bound (Z_{max}) coincides with the value of a feasible solution (Z_{UB}) and so Z_{UB} must be the optimal solution. As can be seen from Figure 6.3, the only place at which a lower bound and an upper bound can coincide on the value line is at the optimal solution.

(2) Set $\pi = 2$ initially. If Z_{max} has not improved (i.e. increased) in the last N subgradient iterations with the current value of π then halve π. A value for N of 30 would seem to be reasonable, although it might well be worthwhile experimenting (computationally) with different values of N, to see whether a different value generates significantly better results for the particular problem under consideration.

(3) Unless terminated by **(1)** above, terminate the subgradient optimization procedure when π is small (e.g. $\pi \leq 0.005$). Often this means that many hundreds of subgradient iterations are done, but this is inevitable—if we want to generate a good quality lower bound (i.e. a value for Z_{max} which is close to the optimal solution) we cannot reasonably expect to do so in just a few subgradient iterations.

(4) In the author's experience, how the Lagrange multipliers are initialized is not particularly important; how quickly the subgradient procedure terminates is relatively insensitive to the initial choice of multipliers, as is the quality of the final lower bound Z_{max} obtained.

(5) It is helpful to replace Z_{UB} in the equation

$$T = \pi(Z_{UB} - Z_{LB}) / \sum_{i=1}^{m} G_i^2$$

for the step size T by $1.05Z_{UB}$.

The reasoning here is that, ideally, we would like Z_{UB} and Z_{LB} to coincide ($Z_{UB} = Z_{LB}$ implies that we have found the optimal solution). As Z_{LB} approaches Z_{UB}, using the original equation for the step size implies that T gets smaller and smaller, so that we may experience many iterations as Z_{LB} creeps closer and closer to Z_{UB}. Replacing Z_{UB} by $1.05Z_{UB}$ (where the 1.05 is arbitrarily chosen) avoids this problem.

(6) Suppose that $\lambda_i = 0$ and $G_i < 0$, then λ_i will remain zero after update. But a G_i^2 factor will been included in the equation for T. There seems little effective reason for this—after all we know that λ_i is not going to alter. Hence we have found it helpful to adjust the subgradients before calculating T using:

$$G_i = 0, \text{ if } \lambda_i = 0 \text{ and } G_i < 0$$

(7) It may happen that, at some subgradient iteration, $G_i = 0, \forall i$ (assuming that the adjustment referred to above has been carried out), and the subgradient procedure terminates. If this happens then:

1. if the current LLBP solution (X_j) is feasible for the original problem then it is the optimal solution for the original problem (i.e. it satisfies the conditions given earlier for the solution to LLBP to be the optimal solution to the original problem);
2. otherwise we simply have a lower bound Z_{max} on the optimal solution to the original problem.

(8) It is important to formulate the problem such that all equations which are to be relaxed are *scaled*, so that their right-hand sides have a maximum absolute value of one (or are zero).

This is because relaxing constraints with varying right-hand sides can lead to a constraint with a very large right-hand side dominating the step size expression for T.

(9) When relaxing an equality constraint then (as mentioned previously) λ_i can be positive or negative and the expression for updating λ_i becomes

$$\lambda_i = \lambda_i + TG_i$$

(10) Observant readers will have noted that nowhere in the subgradient procedure given above did we mention Lagrangean heuristics or problem reduction. Yet we know from our earlier discussion that both of these techniques yield useful information about the optimal solution, both in terms of an upper bound (Lagrangean heuristic) and in terms of problem reduction.

Plainly it is a simple matter to introduce both of these techniques into the subgradient procedure. Specifically, after solving LLBP, first apply the Lagrangean heuristic (in an attempt to update Z_{UB}), and then apply problem reduction.

For instance, had we done this in the subgradient example considered above we would have:

1. found from the Lagrangean heuristic a feasible solution of value 5, which is better than our (heuristically decided) initial upper

bound of 6 and would have enabled us to update (change) Z_{UB} from $Z_{UB} = 6$ to $Z_{UB} = 5$;

2. eliminated column 4 from the problem and identified column 1 as being in the optimal solution. This would have left us with a reduced (SCP) problem involving just two columns and one row to solve.

(11) Subgradient optimization has been much used in the literature in conjunction with Lagrangean relaxation. The reasons for this are essentially twofold:

- it works—subgradient optimization together with Lagrangean relaxation gives good quality lower bounds for a wide variety of combinatorial optimization problems;
- subgradient optimization works from a general description $[\sum_{j=1}^{n} a_{ij}x_j \geq b_i;\ \ i = 1, \ldots, m]$ of the relaxed constraints, and as such is capable of being directly applied, without alteration, whatever the exact nature of those constraints (cf. multiplier adjustment below).

6.5.3 Multiplier adjustment

Multiplier adjustment is simply a heuristic that:

- generates a starting set of Lagrange multipliers;
- attempts to improve them in some systematic way so as to generate an improved lower bound; and
- repeats the procedure if an improvement is made.

Often we simply change a single multiplier at each iteration, in contrast to subgradient optimization where we (potentially) change all multipliers at each iteration.

The advantages of multiplier adjustment are that

- it is usually computationally cheap; and
- it usually produces an increase (or at least no decrease) in the lower bound at each iteration.

The price we pay for this advantage is:

- the final lower bound obtained can be poor (i.e. worse than that obtained from subgradient optimization); and

- different problems require different multiplier adjustment algorithms (unlike subgradient optimization which is capable of being applied directly to many different problems).

Multiplier adjustment is sometimes called *Lagrangean dual ascent* as it can be viewed as an ascent procedure (i.e. a procedure with a monotonic improvement in the lower bound at each iteration) for the Lagrangean dual program.

As in the development of Lagrangean heuristics, developing multiplier adjustment algorithms is an art. However, *exactly* as for subgradient optimization above, where the equation for updating multipliers was:

$$\lambda_i = \max(0, \lambda_i + TG_i); \quad i = 1, \ldots, m$$

the direction in which we would like to change multipliers is clear:

- if $G_i < 0$ we would like to reduce λ_i;

- if $G_i = 0$ we leave λ_i unchanged;

- if $G_i > 0$ we would like to increase λ_i.

Hence one possible (very simple) multiplier adjustment algorithm for the SCP is:

1. solve LLBP with the current set of multipliers (λ_i) ;

2. choose any row i for which $G_i > 0$ (i.e. row i is uncovered in the current LLBP solution);

3. if row i is uncovered then it is easy to see from the relevant mathematics of LLBP that:

   (a) increasing λ_i will increase the lower bound obtained from LLBP; and

(b) the maximum amount (δ) by which we can increase λ_i before the solution to LLBP changes is given by:

$$\delta = \min_{j=1,\dots,n} (C_j | a_{ij} = 1)$$

i.e. $\delta = \min(C_j)$ such that column j covers row i;

4. increase λ_i by δ and go to 1.

The above multiplier adjustment algorithm terminates when all rows are covered (i.e. $G_i \leq 0;\ \forall i$).

To illustrate this multiplier adjustment algorithm we shall apply it to our example SCP, starting from the multiplier values $\lambda_1 = 1.5, \lambda_2 = 1.6$ and $\lambda_3 = 2.2$ as before:

1. the solution to LLBP is $X_1 = 1, X_2 = X_3 = X_4 = 0, Z_{LB} = 4.2$ with $C_1 = -1.1, C_2 = 0.8, C_3 = 0.3, C_4 = 1.2$ and $G_1 = G_2 = 0, G_3 = 1$;

2. row 3 is uncovered as $G_3 > 0$;

3. columns 2, 3 and 4 cover row 3 so

$$\delta = \min(C_2, C_3, C_4) = 0.3;$$

4. we increase λ_3 by 0.3 to give a new set of multipliers $\lambda_1 = 1.5, \lambda_2 = 1.6, \lambda_3 = 2.5$.

Resolving LLBP with this new set of multipliers gives $X_1 = X_3 = 1, X_2 = X_4 = 0$ with a new lower bound of $Z_{LB} = 4.5$. As we would expect (from the manner in which we designed this multiplier adjustment algorithm), the lower bound has increased, and as all rows are now covered ($G_i \leq 0;\ \forall i$ in the LLBP solution associated with $Z_{LB} = 4.5$) the algorithm terminates.

Plainly we could have designed a better multiplier adjustment algorithm, for example investigating not just *increasing* λ_i as above, but also *reducing* λ_i. Discovering whether a particular multiplier adjustment algorithm gives good quality lower bounds at reasonable computational cost is a matter for computational experimentation.

We see here that, as previously remarked, in contrast to subgradient optimization where we simply apply a sequence of general formulae for subgradients, step size and Lagrange multiplier update, multiplier adjustment algorithm design is a much more creative (difficult!) process.

6.6 Dual Ascent

Dual ascent came to prominence with the work of Erlenkotter [11] and Bilde and Krarup [12] on the uncapacitated warehouse location problem which, computationally, was very successful.

6.6.1 Basic concepts

Consider the linear programming (LP) relaxation of any combinatorial optimization problem P (assumed to be a minimization problem, so that the LP relaxation is also a minimization problem).

The dual linear program associated with the LP relaxation is therefore a maximization problem. Hence:

optimal P (integer) solution
$\geq$
LP relaxation solution
$=$
dual LP solution
$\geq$
any feasible solution for the dual LP

Therefore any *heuristic* for the dual LP provides a way of generating a lower bound on the optimal integer solution of the original problem, since *any* dual feasible solution gives a lower bound on the optimal integer solution to the original problem.

Figure 6.4 illustrates the situation. In that figure we essentially have three regions:

1. upper bounds, the region above the optimal (integer) solution;
2. dual LP feasible solutions, the region below the LP relaxation solution; and
3. the gap, the region between the LP relaxation solution and the optimal (integer) solution.

Dual ascent consists therefore of simply thinking up some heuristic for generating feasible solutions to the dual of the LP relaxation of a problem.

We shall illustrate dual ascent with reference to the set covering problem. For the SCP the LP relaxation is:

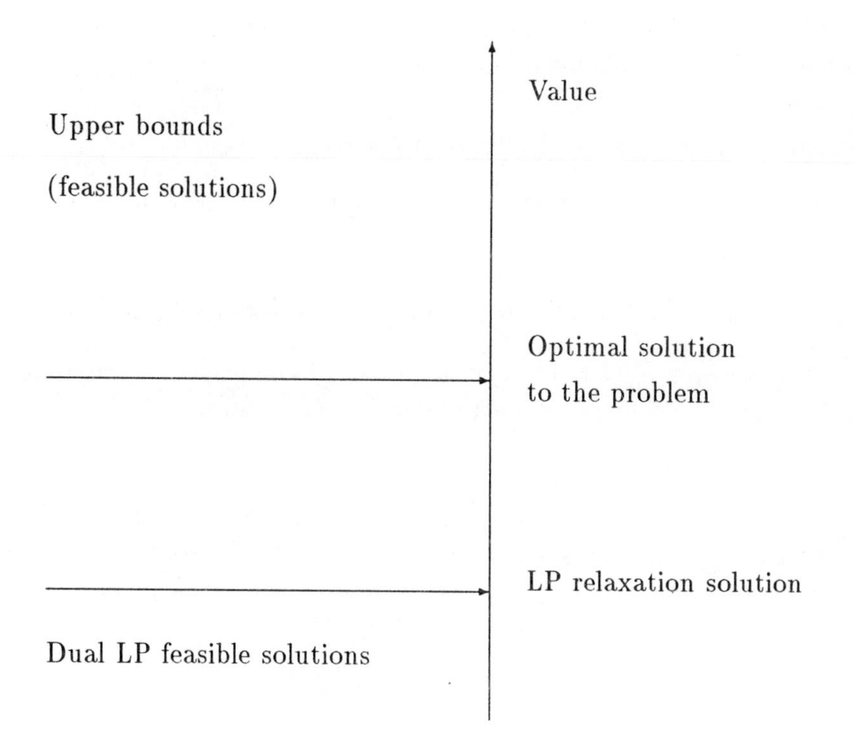

Figure 6.4: Dual ascent

$$\begin{array}{ll} \text{minimize} & \sum_{j=1}^{n} c_j x_j \\ \text{subject to} & \sum_{j=1}^{n} a_{ij} x_j \geq 1; \;\; i = 1, \ldots, m \\ & 0 \leq x_j \leq 1; \;\; j = 1, \ldots, n \end{array}$$

As we have assumed (see above) that all costs c_j are strictly greater than zero this LP relaxation can be written as:

$$\begin{array}{ll} \text{minimize} & \sum_{j=1}^{n} c_j x_j \\ \text{subject to} & \sum_{j=1}^{n} a_{ij} x_j \geq 1; \;\; i = 1, \ldots, m \\ & 0 \leq x_j; \;\; j = 1, \ldots, n \end{array}$$

and the dual LP is:

$$\begin{array}{ll} \text{maximize} & \sum_{i=1}^{m} u_i \\ \text{subject to} & \sum_{i=1}^{m} u_i a_{ij} \leq c_j; \;\; j = 1, \ldots, n \\ & u_i \geq 0; \;\; i = 1, \ldots, m \end{array}$$

In order to illustrate dual ascent we will develop a dual ascent algorithm for our example SCP.

6.6.2 Example dual ascent algorithm

Considering the dual LP given above one possible (very simple) dual ascent algorithm is:

1. set $u_i = 0$; $\forall i$ (this is a feasible solution for the dual LP);

2. take each u_i ($i = 1, \ldots, m$) in turn and increase it by as much as possible consistent with retaining feasibility.

A key point to note here is that often in a dual ascent algorithm we start from some dual feasible point and always retain dual feasibility throughout the algorithm. To illustrate the dual ascent algorithm given above we shall apply it to our example SCP, for which the dual LP is:

$$\begin{array}{ll} \text{maximize} & u_1 + u_2 + u_3 \\ \text{subject to} & u_1 + u_2 \leq 2 \\ & u_3 \leq 3 \\ & u_1 + u_3 \leq 4 \\ & u_2 + u_3 \leq 5 \\ & u_1, u_2, u_3 \geq 0 \end{array}$$

Our simple dual ascent algorithm, as applied to this example, is therefore:

1. set $u_1 = u_2 = u_3 = 0$;

2. (a) after setting $u_2 = u_3 = 0$, the constraints involving u_1 are:
   $u_1 \leq 2$
   $u_1 \leq 4$
   so that u_1 can be increased to 2;

   (b) after setting $u_1 = 2, u_3 = 0$, the constraints involving u_2 are:
   $u_2 \leq 0$
   $u_2 \leq 5$
   so that u_2 cannot be increased;

(c) after setting $u_1 = 2, u_2 = 0$, the constraints involving u_3 are:
$u_3 \leq 3$
$u_3 \leq 2$
$u_3 \leq 5$
so u_3 can be increased to 2.

Hence we have a final solution of $u_1 = 2, u_2 = 0$ and $u_3 = 2$ which is a dual feasible solution and gives a lower bound of $u_1 + u_2 + u_3 = 4$.

Plainly we could have designed a better dual ascent algorithm, for example investigating not just increasing u_i as above, but also reducing u_i (thereby enabling us to increase other u_js). Discovering whether a particular dual ascent algorithm gives good quality lower bounds at reasonable computational cost is a matter for computational experimentation.

6.6.3 Connections

Consider the two techniques for generating lower bounds that we have given above, namely:

- Lagrangean relaxation (with the multipliers being decided by subgradient optimization or multiplier adjustment); and
- dual ascent, i.e. the heuristic solution of the dual of the LP relaxation of the problem.

Can we establish any connection between these two techniques? In fact we can by considering the question:

Is there any relationship between dual variables and Lagrange multipliers?

Recall here that we mentioned before that if a Lagrangean lower bound program (LLBP) had the *integrality property* then the maximum lower bound attainable from LLBP is equal to the value of the LP relaxation of the original problem.

However, it can also be shown that the values of the Lagrange multipliers that maximize the lower bound obtained from LLBP are given by the optimal values for the dual variables in the solution of the LP relaxation of the original problem.

In other words *if* the Lagrangean relaxation has the integrality property *then* the optimal Lagrange multipliers and the optimal dual variables are the same. This immediately implies that the maximum lower bound attainable from any Lagrangean relaxation with the integrality property is equal to the maximum lower bound attainable from any dual ascent algorithm for the problem.

6.6.4 Comparisons

Which of the three techniques—subgradient optimization, multiplier adjustment or dual ascent—should we use for obtaining lower bounds?

In the author's experience, subgradient optimization has always appeared to give very good lower bounds, in particular lower bounds often close to the optimal integer solution (so presumably also close to the maximum theoretically obtainable from the Lagrangean relaxation). Hence the author has never been very keen on multiplier adjustment methods although they appear useful for some problems, e.g. the generalized assignment problem [13, 14]. The only time the author tried dual ascent (for the p-median problem), it was a miserable failure!

There is a deeper point here. Techniques such as subgradient optimization, multiplier adjustment and dual ascent are *potentially* of wide applicability, i.e. they can be applied to a wide range of problems. However these techniques may *fail* computationally when applied across a wide range of problems, and may only be successful (possibly outstandingly successful) on one or two problems.

Based on this point then, to choose between these three lower bound techniques, the author's advice would be:

- subgradient optimization will nearly always work;
- multiplier adjustment may work;
- dual ascent will probably not work.

This advice is supported by evidence from the literature. Looking at the 'best' (most successful) optimal algorithm for a number of different problems, the reader will find that it is usually based on subgradient optimization and only occasionally based on multiplier adjustment or dual ascent.

6.6.5 Conclusions

We have now presented sufficient material for the diligent reader to be able to take a problem and develop lower bounds for it via Lagrangean relaxation. Applying subgradient optimization or multiplier adjustment will hopefully yield a good (close to integer optimal) lower bound.

Heuristic solutions

If we are simply interested in a heuristic solution for the problem we are considering, then the use of a Lagrangean heuristic throughout the subgradient procedure implies that at the end of that procedure:

- Z_{UB} is the value of the best feasible solution found;
- Z_{max} is the value of the best lower bound (on the optimal solution to the original problem) found; and
- the gap between Z_{UB} and Z_{max} (in percentage terms $100(Z_{UB}-Z_{max})/Z_{max}$) gives an indication of the quality of the solution (Z_{UB}) obtained from the Lagrangean heuristic.

In other words, *after* having solved the problem we find that we have a solution for which we have an associated *quality guarantee*. This fact is used extensively in many of the papers presented in the literature (see the discussion in section 6.8.).

However, we may be able to go further than simply accepting whatever quality guarantee figure we end up with after the subgradient procedure (note that we have no control or influence over this figure). This can be done by specifying, *before* we start to solve the problem, what quality guarantee we would like to achieve:

- suppose we are prepared to accept a solution within $\alpha\%$ of the optimum Z_{OPT}, i.e. we would accept a solution Z_{UB} with $(Z_{UB} - Z_{OPT}) \leq (\alpha/100)Z_{OPT}$, then
- *instead* of using Z_{UB} in any problem reduction test (for fixing variables to zero or one) use $Z_{UB}/[1 + (\alpha/100)]$.

If this is done then the subgradient procedure will terminate with either:

1. Z_{UB} proved to be within $\alpha\%$ of optimal, so that we will have achieved our desired (pre-specified) quality guarantee; or

2. a reduced problem, the optimal solution ZR_{OPT} of which has $\min(ZR_{OPT}, Z_{UB})$ within $\alpha\%$ of optimal. In this case we will not have achieved our pre-specified quality guarantee of $\alpha\%$ but will instead have a solution Z_{UB} with an associated quality guarantee of $100(Z_{UB} - Z_{max})/Z_{max}$ $(> \alpha\%)$.

Optimal solutions

If we are interested in finding the optimal solution to the problem we are considering then:

- we may already have found the optimal solution ($Z_{max} = Z_{UB}$); but

- if not we need to resort to a tree search (branch-and-bound) procedure to resolve the problem.

Tree search procedures are of course not heuristics—at least not in the sense in which we use the term in this book. (As has been pointed out in Chapter 1, some authors do use the word 'heuristic' for such methods, but it is not the consensus usage.) Nevertheless, for the sake of completeness, and for the benefit of readers who may wish to prove optimality in a particular situation, we consider tree search procedures below.

6.7 Tree Search

There are a number of different (and difficult) decisions that must be faced in designing a tree search procedure for a particular problem. In this section we concentrate upon designing a tree search procedure using a lower bound based upon Lagrangean relaxation and subgradient optimization. However, much of what is given here is equally applicable to tree search procedures using lower bounds based upon linear programming relaxation, dual ascent or multiplier adjustment.

6.7.1 Tree structure

In the author's experience, a binary tree search has always worked in every case on which it was tried.

6.7.2 Branching node

The author always uses a depth-first strategy (i.e. branch from the last available node in the tree). Again, this has always worked in all cases tried. The reader would only be recommended to try a more complex tree search if in complete despair at the results from a binary depth-first tree search.

6.7.3 Backtracking

We can backtrack from a tree node if either:

- the tree node is infeasible; or
- the lower bound at the tree node exceeds the upper bound.

Note here that the fact that the solution to the Lagrangean lower bound program (LLBP) at a particular tree node is a feasible solution to the original problem is *not* a condition for backtracking (see the fallacy mentioned above in section 6.3.5).

6.7.4 Forward branching rule

The author has always used a simple rule for forward branching based upon choosing a single variable and setting that variable equal to one (assuming here that we are dealing with a mathematical program involving zero-one variables).

Hence in a tree search procedure based upon a linear programming (LP) bound, for example, a reasonable (and simple) rule to use is: choose the variable that has the largest fractional part in the LP solution at the current tree node and branch by setting that variable to one.

As we are considering a tree search based upon Lagrangean relaxation, things are more complicated than this. Recall the LLBP:

$$\begin{aligned} \text{minimize} \quad & cx + \lambda(b - Ax) \\ & = (c - \lambda A)x + \lambda b \\ \text{subject to} \quad & Bx \geq d \\ & x \in (0, 1) \end{aligned}$$

In a tree search procedure based upon a Lagrangean relaxation bound a reasonable rule to use is as follows:

1. Let (λ, X) represent the multipliers and variable values associated with the best lower bound found at the tree node from which we are branching.

2. Consider the vector of values $\lambda(b - AX) = \lambda G$ and choose the constraint (i, say) with the maximum (absolute) value in this vector.

   As discussed above, if X is feasible and satisfies $\lambda(b - AX) = 0$ then it is the optimal solution to the problem. Choosing the constraint with the maximum (absolute) value in $\lambda(b - AX)$ and then, by branching, attempting to satisfy that constraint (thereby reducing the contribution of constraint i to $\lambda(b - AX)$, i.e. attempting to make $\lambda(b - AX)$ zero) would seem logical.

3. Define:

   $$S = \{j | x_j \text{ appears in constraint } i, X_j = 1\};$$

   if $S = \emptyset$ (i.e. no variables in constraint i are in the current LLBP solution X_j) then define:

   $$S = \{j | x_j \text{ appears in constraint } i, X_j = 0\}$$

   Choose the variable to set to one that has the minimum LLBP objective function coefficient ($C' = c - \lambda A$) out of those variables in S (i.e. the one that is most 'likely' to be one in an optimal completion of the current tree node).

Obviously, when choosing a variable to branch on at a particular tree node we exclude from consideration any variables already fixed to zero or one (either by explicit branching or by problem reduction at previous tree nodes).

When programming a tree search based upon Lagrangean relaxation, the author's advice would be to try the above branching rule, and only investigate different branching rules if the computational results from the above rule are not satisfactory.

To illustrate the above branching rule, consider our example LLBP solution as given earlier with $Z_{LB} = 4.2$. Which variable would we choose to branch on?

1. We have $\lambda_1 = 1.5, \lambda_2 = 1.6, \lambda_3 = 2.2, X_1 = 1, X_2 = X_3 = X_4 = 0$. We also know that $C_1 = -1.1, C_2 = 0.8, C_3 = 0.3, C_4 = 1.2$ from above.

2. Then:

$$\begin{aligned} \lambda_1(1 - X_1 - X_3) &= 0 \\ \lambda_2(1 - X_1 - X_4) &= 0 \\ \lambda_3(1 - X_2 - X_3 - X_4) &= 2.2 \end{aligned}$$

   Hence we would choose constraint 3. Note that this constraint corresponds to an uncovered row in the current LLBP solution $(X_1 = 1, X_2 = X_3 = X_4 = 0)$.

3. $S = \{j | x_j \text{ appears in constraint } 3, X_j = 1\}$ gives $S = \emptyset$ since only columns (variables) 2, 3 and 4 appear in constraint 3 and they are all currently zero.
   Hence $S = \{j | x_j \text{ appears in constraint } 3, X_j = 0\}$ which gives $S = \{2, 3, 4\}$ and these variables have LLBP objective function coefficients of $C_2 = 0.8, C_3 = 0.3, C_4 = 1.2$. The minimum of these values is $C_3 = 0.3$, so we would choose x_3 to set to one in the forward branch.

6.7.5 Lower bound computation

Initial tree node

At the initial tree node (before branching), it is the author's experience that it is well worth while expending computational effort in doing many (of the order of hundreds) iterations.

This gives the Lagrangean heuristic a chance to find a good feasible solution, the problem reduction tests a chance to reduce the size of the problem and the subgradient procedure the chance to find a good lower bound. Given that hard problem instances, by definition, require many tree nodes for solution, the time spent at the initial tree node soon becomes a small fraction of the overall computation time for such problems.

Other tree nodes

One option is to simply repeat, at each tree node, *exactly* the same procedure as at the initial tree node. The disadvantage of this is that it can make the lower bound computation at each tree node rather expensive. Essentially we have to balance the computational effort involved in calculating the lower bound at each tree node against the potential saving in tree nodes resulting from improving that lower bound.

Nowadays the author tends to use the strategy of:

1. initialising the Lagrange multipliers at each tree node with the set associated with the best (maximum) lower bound found at the predecessor tree node;

2. carrying out 30 subgradient iterations, with π (the user-defined parameter satisfying $0 < \pi \leq 2$) being halved every 5 (or 10) iterations at tree nodes associated with forward branching;

3. doubling both these figures (to 60 and 10 (or 20) respectively) for tree nodes associated with backtracking.

The logic for step 3 above is that in backtracking we are explicitly setting to zero in the tree a variable which, when we carried out the forward branch, was a good candidate for taking the value one (cf. the rule for forward branching given above). It seems reasonable therefore to do more iterations at the backtrack tree node in an attempt to achieve a lower bound sufficient to eliminate the node from further consideration.

6.7.6 Computer programs

Clearly a number of readers of this chapter and book are likely to be doctoral students who are not only gaining additional knowledge, but may also decide to apply a number of the techniques given in this chapter to a particular research problem. For this reason we feel that some guidance/help should be given about sources of available computer programs. The help that we can offer the reader falls into two categories:

- solving the Lagrangean relaxation;
- problem reduction.

We deal with each in turn.

Lagrangean relaxation

Recall here that one of the guidelines for choosing a Lagrangean relaxation was that the relaxed problem (the Lagrangean lower bound program) was, computationally, easy to solve. Often the LLBP is a standard problem, for example a knapsack problem.

Generally it is best (if at all possible) to use commercial/publically available software to solve these standard problems rather than writing one's own software (software development is both time-consuming and frustrating!). An excellent source of information on available commercial software is contained in the advertisements/software reviews in *OR/MS Today* (the magazine associated with membership of the Operations Research Society of America).

Table 6.1 gives brief details of additional useful software.

Problem reduction

Note here that, in any program code, it is often difficult to alter data structures to reflect explicitly the fact that variables have been eliminated from the problem (either by being fixed to zero or by being fixed to one). There are some specific ways to avoid altering data structures:

- If a variable x_k is fixed to zero then simply alter its cost c_k from its original value to infinity (in computer terms a very large number). This will ensure that the variable never appears in a solution to the LLBP (nor, presumably, in any solution found by the Lagrangean heuristic).
- If a variable x_k is fixed to one then:

Table 6.1: Useful software

Linear/integer programming There are many linear programming packages available, often with extensions to allow mixed-integer (or pure integer) problems to be solved by LP based tree search. One worth noting is the Marsten XMP code which is unusual in that one has complete access to the FORTRAN source (contact XMP Software, PO Box 58119, Tucson AZ 85732, USA).
Knapsack problem The book by Martello and Toth, *Knapsack Problems: Algorithms and Computer Implementations* (Wiley, 1990) contains a number of codes.
Generalized assignment problem A code is available from Guignard and Rosenwein [14]
Assignment/transportation/network flow problems A code is available from Bertsekas and Tseng (Operations Research **36** (1988), 93-114).
Miscellaneous A number of codes are available from ORSEP (Operations Research Software Exchange Programme) - see issues of the European Journal of Operational Research. Packages (such as QSB+, STORM), capable of solving relatively small problems on IBM-PCs, are readily available, e.g. see the review by Beasley (Journal of the Operational Research Society **39** (1988), 487-509. There are several books which contain useful codes e.g. Syslo, Deo and Kowalik, *Discrete Optimization Algorithms: with Pascal programs* (Prentice-Hall, 1983). The NAG library (contact NAG, Wilkinson House, Jordan Hill Road, Oxford OX2 8DR, England) contains a number of useful codes.

1. any relaxed constraint i which is always satisfied because $x_k = 1$ can be:
   (a) neglected in LLBP simply by always setting $\lambda_i = 0$;
   (b) explicitly neglected in any Lagrangean heuristic;
2. accumulate a total C_{add}, which is the total cost of variables identified as being in the solution, and:
   (a) alter the cost c_k of x_k from its original value to a value v_k, say, sufficient to ensure that x_k will always automatically appear in the solution to LLBP (for the SCP example given above $v_k = 0$ would be appropriate);
   (b) add C_{add} minus the sum of all the v_k to the solution value given by LLBP to get a valid lower bound;
   (c) ensure that all x_k are included in the Lagrangean heuristic solution procedure.

Our experience has been that, after a significant amount of reduction has been achieved, e.g. 10% of variables fixed at zero or one, it is computationally worthwhile to perform 'house cleaning' operations, i.e. explicitly to alter data structures to reflect the fact that variables have been eliminated from the problem (either by being fixed to zero or by being fixed to one).

The reasoning here is clear: the elimination of just one variable is probably not sufficient to justify the computational overhead of altering data structures, whereas after a significant amount of reduction it is computationally worthwhile altering data structures. Essentially we have to balance the computational effort involved in altering the data structure against the possible reduction in overall computational effort in future iterations resulting from an altered data structure.

In fact, although we have not mentioned it above, this balance between the computational effort involved in some procedure and the future benefits (computational reduction) from the procedure is a key consideration in effective algorithm design.

For example, should we carry out problem reduction (a computational procedure) at each subgradient iteration or only at selected subgradient iterations (e.g. at iterations when Z_{max} increases)? Answering such a question is a matter of specific computational experimentation, i.e. trying different strategies and seeing which is best.

However it is important to be aware that such a question needs to be addressed.

6.8 Applications

As mentioned at the very beginning of this chapter much of what is contained in it is historical, and has been known for several years. To enable the reader to get some insight into current up-to-date work involving Lagrangean relaxation we have, in this section, adopted the strategy of:

- selecting calendar year 1991 (the most recent complete calendar year at the time of writing);
- selecting a number of journals; and
- reviewing all papers in those journals in our selected year which use Lagrangean relaxation.

This provides *systematic* insight into the range of problems to which Lagrangean relaxation is being applied.

This approach is not common in chapters of this kind. What is usually done is for the author to present a few 'selected highlights', i.e. papers drawn from a number of years that are meant to be especially significant. We have adopted a different approach because we believe that Lagrangean relaxation has wide applicability, and that a systematic survey (rather than 'selected highlights') provides *evidence* for such a belief [1].

By the end of this chapter readers will be able to judge for themselves whether Lagrangean relaxation has wide applicability or not.

6.8.1 Journals

The journals that we have selected for systematic review are:

- *Operations Research*;

[1] *Editorial footnote*: Readers will notice that the alternative strategy is indeed that adopted in the main by other contributors to this book. It is perhaps because Lagrangean relaxation is (relatively speaking) so much more developed than the other techniques presented that Beasley's approach is possible.

- *Management Science*;
- *European Journal of Operational Research.*

These journals are some of the most prestigious and most cited journals in the field of Operations Research/Management Science.

Operations Research

In 1991 *Operations Research* published 86 papers (of assorted lengths) spread over 1026 journal pages. Of these, 6 papers, spread over 76 pages, (7.0% and 7.4% respectively) involved Lagrangean relaxation. These papers are dealt with below.

- Ahmadi and Matsuo [15] present a paper dealing with the line segmentation problem. This is the problem of allocating the machines in a multi-stage production line to a number of different 'families' of items.

  They present a quadratic integer programming formulation of the problem and use Lagrangean relaxation, subgradient optimization and a Lagrangean heuristic. Computational results are presented for problems based on real-world data for large-scale circuit board manufacturing.

- Balas and Saltzman [16] present a paper dealing with the three-index assignment problem. This is a generalization of the well-known (two-index) assignment problem (involving variables x_{ij} indicating whether i and j are assigned together) to a three-index problem (involving variables x_{ijk} indicating whether i, j and k are assigned together).

  They present a zero-one formulation of the problem and use Lagrangean relaxation, subgradient optimization, dual ascent and a Lagrangean heuristic in conjunction with a tree search procedure. Computational results are given for randomly generated problems.

- Shulman [17] presents a paper dealing with the dynamic capacitated plant location problem. This is the problem of deciding the size, location and opening period of plants so as to service customers at minimum cost over a specified time horizon.

He presents a mixed-integer programming formulation of the problem and uses Lagrangean relaxation, subgradient optimization and Lagrangean heuristics. Computational results are presented for a number of randomly generated problems.

- Altinkemer and Gavish [18] present a paper dealing with the delivery problem. This is the problem of deciding minimum cost routes for a fleet of delivery vehicles (of known capacity) which are based at a central depot and have to visit a set of customers.

  They present a zero-one formulation of the problem and give a number of heuristics for the problem. A lower bound, based upon Lagrangean relaxation and subgradient optimization, is used to evaluate the quality of the heuristic solutions. Computational results are presented for a number of problems (both taken from the literature and randomly generated).

- Noon and Bean [19] present a paper dealing with the asymmetric generalized travelling salesman problem. This is a generalization of the well-known travelling salesman problem in which there are sets of cities, and the tour must include exactly one city from each of the (mutually exclusive) sets.

  They present a zero-one formulation of the problem and use Lagrangean relaxation, subgradient optimization, problem reduction and a Lagrangean heuristic in conjunction with a tree search procedure. Computational results are presented for a number of randomly generated problems.

- Ahmadi and Tang [20] present a paper dealing with the operation partitioning problem. This is the problem of assigning operations to machines so as to minimize the total movement of jobs between machines.

  They present two zero-one formulations of the problem and use Lagrangean relaxation, subgradient optimization and Lagrangean heuristics. Computational results are presented for a number of randomly generated problems.

Management Science

In 1991 *Management Science* published 115 papers spread over 1652 journal pages. Of these 6 papers, spread over 109 journal pages, (5.2% and 6.6% respectively) involved Lagrangean relaxation. These papers are dealt with below.

- Pirkul and Schilling [21] present a paper dealing with the capacitated maximal covering problem. This is the problem of locating a certain number of facilities to cover (service) selected demand points, where facilities are both limited in size and in the total demand that they can service. This problem is related to locating emergency service facilities, such as fire and ambulance stations.

  They use a zero-one formulation of the problem and apply Lagrangean relaxation with subgradient optimization and a Lagrangean heuristic. Computational results are presented for a large number of problems, including a 'real world' problem involving locating ten fire stations in an area of 85 square miles involving 625 demand points.

- Campbell and Mabert [22] present a paper dealing with planning production of a number of items on a single machine. The key feature of this problem is that production is cyclical, i.e. for each item the time between each lot (batch) being produced is a constant.

  They present a mixed-integer programming formulation of the problem and use Lagrangean relaxation, subgradient optimization and a Lagrangean heuristic. Computational results are presented for a number of problems, including some based on data from the Ford Motor company.

- Gavish and Pirkul [23] present a paper dealing with the multi-resource generalized assignment problem. This is the problem of minimising the total cost involved in assigning tasks to agents, where associated with each possible task/agent assignment is a vector of resources and there are limits on the resources available to each agent.

  They present a zero-one formulation of the problem and apply Lagrangean relaxation, subgradient optimization, Lagrangean heuristics and problem reduction in conjunction with a tree search procedure. Computational results are presented for a number of randomly generated problems.

- Domich *et al.* [24] present a paper dealing with the problem of locating offices for the American Internal Revenue Service (IRS). This problem can be formulated as an uncapacitated fixed-charge location-allocation problem.

  They use a zero-one formulation of the problem and apply Lagrangean relaxation with subgradient optimization and a Lagrangean heuristic. Computational results are presented for real-world data based upon locating IRS offices in Florida.

- Bard and Bejjani [25] present a paper dealing with the problem of leasing telecommunication lines so as to achieve a desired level of service at minimum cost.

  They present a dynamic programming formulation of the problem which involves solving a number of general integer programs. These integer programs are solved using Lagrangean relaxation, subgradient optimization and a Lagrangean heuristic. Computational results are presented for real-world data involving a telecommunications company based in Texas.

- Drexl [26] presents a paper dealing with resource constrained project scheduling. This is the problem of minimising the cost of performing a set of jobs subject to job precedence constraints and resource restrictions.

  He presents a zero-one formulation of the problem and uses a number of bounds (including a bound based on Lagrangean relaxation and multiplier adjustment) in conjunction with a tree search procedure. Computational results are presented for a number of problems (both taken from the literature and randomly generated).

European Journal of Operational Research

In 1991 the *European Journal of Operational Research* published 207 papers spread over 2342 journal pages. Of these 9 papers, spread over 108 pages, (4.3% and 4.6% respectively) involved Lagrangean relaxation. These papers are dealt with below.

- Cornuejols *et al.* [27] present a paper dealing with the capacitated plant location problem. This is the problem of deciding the number, and location, of plants to service customer demands, at minimum cost, where there is a limit on the total demand that can be serviced from each plant.

  They present a mixed-integer programming formulation of the problem and consider a number of different Lagrangean relaxations. They use subgradient optimization and a Lagrangean heuristic. Computational results are presented for a number of problems (both taken from the literature and randomly generated).

- Aboudi *et al.* [28] present a paper dealing with the well-known assignment problem but with additional constraints involving equality between certain solution variables.

  They present a zero-one formulation of the problem and use Lagrangean relaxation and a procedure for generating valid inequalities violated by Lagrangean relaxation solutions. No computational results are given (only a small numerical example).

- Sharma [29] presents a paper dealing with a problem involving fertiliser distribution in India. This is a problem of deciding how much fertiliser to produce at each manufacturing plant, which warehouses to use to store it (and their sizes), and how to distribute fertiliser to customers over a specified time horizon.

  He presents a mixed-integer programming formulation of the problem and uses Lagrangean relaxation, subgradient optimization and a Lagrangean heuristic. Computational results are presented for problems based upon real-world data.

- Deckro *et al.* [30] present a paper dealing with the multi-project resource constrained scheduling problem. This is the problem of minimizing the cost of performing a set of jobs (spread across a number of projects) subject to job precedence constraints and resource restrictions.

  They present a zero-one formulation of the problem and use Lagrangean relaxation. Computational results are presented for a number of randomly generated problems.

- Lozano *et al.* [31] present a paper dealing with the single level capacitated dynamic lot-sizing problem. This is the problem of determining the quantities and timing of production batches in order to meet customer demands at minimum cost over a specified time horizon.

  They present a mixed-integer programming formulation of the problem and use Lagrangean relaxation. Computational results are presented for a number of problems (both taken from the literature and randomly generated).

- Wright and von Lanzenauer [32] present a paper dealing with the fixed-charge problem. This is a linear programming problem with discontinuities in the objective function resulting from fixed charges associated with non-zero variable values.

  They present a mixed-integer programming formulation of the problem and use Lagrangean relaxation, multiplier adjustment and a Lagrangean heuristic. Computational results are presented for a number of problems taken from the literature.

- Current and Pirkul [33] present a paper dealing with the problem of designing a minimum cost two-level network, consisting of a primary path between an origin node and a destination node, with all nodes not on the path being connected to the path. In addition trans-shipment facilities must be located on the primary path.

  They present a zero-one formulation of the problem and use Lagrangean relaxation, subgradient optimization and Lagrangean heuristics. Computational results are presented for a number of randomly generated problems.

- Barcelo *et al.* [34] present a paper dealing with the capacitated plant location problem. This is the problem of deciding the number, and location, of plants to service customer demands, at minimum cost, where there is a limit on the total demand that can be serviced from each plant.

  They present a mixed-integer programming formulation of the problem and use Lagrangean relaxation, Lagrangean decomposition, subgradient optimization, a Lagrangean heuristic and problem reduction in conjunction with a tree search procedure. Computational results are presented for a number of randomly generated problems.

- Millar and Gunn [35] present a paper dealing with trawler fleet dispatching. This is the problem of finding a minimum cost plan for using fishing trawlers in order to meet demand for various species of fish at processing plants over a specified time horizon.

  They present a mixed-integer programming formulation of the problem and use Lagrangean relaxation and subgradient optimization. Computational results are presented for real-world data involving a large fish-processing company based in Canada.

6.8.2 Summary

Table 6.2: Summary of 1991 papers reviewed

Authors	Subgr. opt.	Mult. adj.	Dual ascent	Lagr. heur.	Prob. red.	Tree search
Ahmadi/Matsuo [15]	yes	-	-	yes	-	-
Balas/Saltzman [16]	yes	-	yes	yes	-	yes
Shulman [17]	yes	-	-	yes	-	-
Altinkemer/Gavish [18]	yes	-	-	-	-	-
Noon/Bean [19]	yes	-	-	yes	yes	yes
Ahmadi/Tang [20]	yes	-	-	yes	-	-
Pirkul/Schilling [21]	yes	-	-	yes	-	-
Campbell/Mabert [22]	yes	-	-	yes	-	-
Gavish/Pirkul [23]	yes	-	-	yes	yes	yes
Domich *et al.* [24]	yes	-	-	yes	-	-
Bard/Bejjani [25]	yes	-	-	yes	-	-
Drexl [26]	-	yes	-	-	-	yes
Cornuejols *et al.* [27]	yes	-	-	yes	-	-
Aboudi *et al.* [28]	-	-	-	-	-	-
Sharma [29]	yes	-	-	yes	-	-
Deckro *et al.* [30]	-	-	-	-	-	-
Lozano *et al.* [31]	-	-	-	-	-	-
Wright/von Lanzenauer [32]	-	yes	-	yes	-	-
Current/Pirkul [33]	yes	-	-	yes	-	-
Barcelo *et al.* [34]	yes	-	-	yes	yes	yes
Millar/Gunn [35]	yes	-	-	-	-	-

Table 6.2 provides a summary of the papers considered above. The popularity of subgradient optimization and Lagrangean heuristics is clear from this table. In particular note that a common theme nowadays is to use a Lagrangean heuristic to generate a feasible solution to a problem (thereby automatically obtaining an associated quality guarantee), without going to the computational expense of solving the problem optimally.

6.9 Conclusions

In this chapter we have discussed:

- Lagrangean relaxation
- Lagrangean heuristics
- problem reduction
- subgradient optimization
- multiplier adjustment
- dual ascent
- tree search

Plainly there are many possible techniques that can be brought to bear on combinatorial optimization problems in an attempt to develop successful algorithms. In this chapter we have outlined a number of the more popular techniques.

As mentioned before, developing algorithms which are computationally successful at solving combinatorial optimization problems is an art. However, the basic building blocks are clear—the types of techniques considered in this and other chapters of this book.

In many ways the work that goes on (all around the world) developing algorithms for the solution of combinatorial optimization problems is analogous to a large group of children playing with plastic bricks in an attempt to build a house. All children have the same set of bricks to play with but some are more successful than others at building a house. Some, indeed, enjoy the process more than others.

One of the appeals of solving combinatorial optimization problems, at least for this author, is that only a small amount of knowledge is required in order to join in the game with the other children and, through experience, one learns to build better houses with the building blocks at one's disposal.

Further references

The list of references on the following pages also includes additional reading as follows:

- for the set covering problem see [4, 7, 36, 38, 41];
- for Lagrangean relaxation see [39, 40, 43];
- for subgradient optimization see [37, 39, 40, 43, 44, 49];
- for multiplier adjustment see [39, 40, 45, 48];
- for dual ascent see [11, 12, 16, 42];
- for Lagrangean decomposition see [46, 47, 50, 51].

References

[1] M.Held and R.M.Karp (1970) The travelling-salesman problem and minimum spanning trees. *Ops.Res.*, **18**, 1138-1162.

[2] M.Held and R.M.Karp (1971) The travelling-salesman problem and minimum spanning trees: Part II. *Math.Prog.*, **1**, 6-25.

[3] J.E.Beasley (1985) A note on solving large p-median problems. *EJOR*, **21**, 270-273.

[4] J.E.Beasley (1987) An algorithm for set covering problems. *EJOR*, **31**, 85-93.

[5] J.E.Beasley (1988) An algorithm for solving large capacitated warehouse location problems. *EJOR*, **33**, 314-325.

[6] J.E.Beasley (1989) An SST-based algorithm for the Steiner problem in graphs. *Networks*, **19**, 1-16.

[7] J.E.Beasley (1990) A Lagrangean heuristic for set covering problems. *NRL*, **37**, 151-164.

[8] J.E.Beasley (1992) Lagrangean heuristics for location problems. *EJOR*, (to appear).

[9] N.Christofides and J.E.Beasley (1982) A tree search algorithm for the p-median problem. *EJOR*, **10**, 196-204.

[10] N.Christofides and J.E.Beasley (1983) Extensions to a Lagrangean relaxation approach for the capacitated warehouse location problem. *EJOR*, **12**, 19-28.

[11] D.Erlenkotter (1978) A dual-based procedure for uncapacitated facility location. *Ops.Res.*, **26**, 992-1009.

[12] O.Bilde and J.Krarup (1977) Sharp lower bounds and efficient algorithms for the simple plant location problem. *Annals of Discrete Math.*, **1**, 79-88.

[13] M.L.Fisher, R.Jaikumar and L.N.Van Wassenhove (1986) A multiplier adjustment method for the generalized assignment problem. *Man.Sci.*, **32**, 1095-1103.

[14] M.Guignard and M.B.Rosenwein (1989) An improved dual based algorithm for the generalized assignment problem. *Ops.Res.*, **37**, 658-663.

[15] R.H.Ahmadi and H.Matsuo (1991) The line segmentation problem. *Ops.Res.*, **39**, 42-55.

[16] E.Balas and M.J.Saltzman (1991) An algorithm for the three-index assignment problem. *Ops.Res.*, **39**, 150-161.

[17] A.Shulman (1991) An algorithm for solving dynamic capacitated plant location problems with discrete expansion sizes. *Ops.Res.*, **39**, 423-436.

[18] K.Altinkemer and B.Gavish (1991) Parallel savings-based heuristics for the delivery problem. *Ops.Res.*, **39**, 456-469.

[19] C.E.Noon and J.C.Bean (1991) A Lagrangean based approach for the asymmetric generalized travelling salesman problem. *Ops.Res.*, **39**, 623-632.

[20] R.H.Ahmadi and C.S.Tang (1991) An operation partitioning problem for automated assembly system design. *Ops.Res.*, **39**, 824-835.

[21] H.Pirkul and D.A.Schilling (1991) The maximal covering location problem with capacities on total workload. *Man.Sci.*, **37**, 233-248.

[22] G.M.Campbell and V.A.Mabert (1991) Cyclical schedules for capacitated lot sizing with dynamic demands. *Man.Sci.*, **37**, 409-427.

[23] B.Gavish and H.Pirkul (1991) Algorithms for the multi-resource generalized assignment problem. *Man.Sci.*, **37**, 695-713.

[24] P.D.Domich, K.L.Hoffman, R.H.F.Jackson and M.A.McClain (1991) Locating tax facilities: a graphics-based microcomputer optimization model. *Man.Sci.*, **37**, 960-979.

[25] J.F.Bard and W.A.Bejjani (1991) Designing telecommunications networks for the reseller market. *Man.Sci.*, **37**, 1125-1146.

[26] A.Drexl (1991) Scheduling of project networks by job assignment. *Man.Sci.*, **37**, 1590-1602.

[27] G.Cornuejols, R.Sridharan and J.M.Thizy (1991) A comparison of heuristics and relaxations for the capacitated plant location problem. *EJOR*, **50**, 280-297.

[28] R.Aboudi, A.Hallefjord and K.Jörnsten (1991) A facet generation and relaxation technique applied to an assignment problem with side constraints. *EJOR*, **50**, 335-344.

[29] R.R.K.Sharma (1991) Modelling a fertiliser distribution system. *EJOR*, **51**, 24-34.

[30] R.F.Deckro, E.P.Winkofsky, J.E.Hebert and R.Gagnon (1991) A decomposition approach to multi-project scheduling. *EJOR*, **51**, 110-118.

[31] S.Lozano, J.Larraneta and L.Onieva (1991) Primal-dual approach to the single level capacitated lot-sizing problem. *EJOR*, **51**, 354-366.

[32] D.Wright and C.H.von Lanzenauer (1991) COAL: a new heuristic approach for solving the fixed charge problem—computational results. *EJOR*, **52**, 235-246.

[33] J.Current and H.Pirkul (1991) The hierarchical network design problem with trans-shipment facilities. *EJOR*, **52**, 338-347.

[34] J.Barcelo, E.Fernandez and K.O.Jörnsten (1991) Computational results from a new Lagrangean relaxation algorithm for the capacitated plant location problem. *EJOR*, **53**, 38-45.

[35] H.H.Millar and E.A.Gunn (1991) Dispatching a fishing trawler fleet in the Canadian Atlantic groundfish industry. *EJOR*, **55**, 148-164.

[36] E.Balas and A.Ho (1980) Set covering algorithms using cutting planes, heuristics, and subgradient optimization: a computational study. *Math.Prog.*, **12**, 37-60.

[37] M.S.Bazaraa and J.J.Goode (1979) A survey of various tactics for generating Lagrangean multipliers in the context of Lagrangean duality. *EJOR*, **3**, 322-338.

[38] T.J.Chan and C.A.Yano (1992) A multiplier adjustment approach for the set partitioning problem. *Ops.Res.*, **40**, S40-S47.

[39] M.L.Fisher (1981) The Lagrangean relaxation method for solving integer programming problems. *Man.Sci.*, **27**, 1-18.

[40] M.L.Fisher (1985) An applications oriented guide to Lagrangean relaxation. *Interfaces*, **15**, 10-21.

[41] M.L.Fisher and P.Kedia (1990) Optimal solution of set covering/partitioning problems using dual heuristics. *Man.Sci.*, **36**, 674-688.

[42] R.D.Galvao (1980) A dual-bounded algorithm for the p-median problem. *Ops.Res.*, **28**, 1112-1121.

[43] A.M.Geoffrion (1974) Lagrangean relaxation for integer programming. *Math.Prog. Study*, **2**, 82-114.

[44] J.L.Goffin (1977) On the convergence rates of subgradient optimization methods. *Math.Prog.*, **13**, 329-347.

[45] M.Guignard (1988) A Lagrangean dual ascent algorithm for simple plant location problems. *EJOR*, **35**, 193-200.

[46] M.Guignard and S.Kim (1987) Lagrangean decomposition: a model yielding stronger Lagrangean bounds. *Math.Prog.*, **39**, 215-228.

[47] M.Guignard and S.Kim (1987) Lagrangean decomposition for integer programming: theory and applications. *RAIRO*, **21**, 307-323.

[48] M.Guignard and M.B.Rosenwein (1989) An application-oriented guide for designing Lagrangean dual ascent algorithms. *EJOR*, **43**, 197-205.

[49] M.H.Held, P.Wolfe and H.D.Crowder (1974) Validation of subgradient optimization. *Math.Prog.*, **6**, 62-88.

[50] K.Jörnsten and M.Nasberg (1986) A new Lagrangean relaxation approach to the generalized assignment problem. *EJOR*, **27**, 313-323.

[51] H.Reinoso and N.Maculan (1992) Lagrangean decomposition in integer linear programming: a new scheme. *INFOR*, **30**, 1-5.

Chapter 7

Evaluation of Heuristic Performance

Colin R Reeves

7.1 Introduction

The earlier chapters of this book have described some of the most encouraging recent developments in heuristic techniques for combinatorial optimization. Many of the papers reviewed in these pages have reported excellent performance for these methods on a wide variety of problem types. However, there are clearly some questions that still await answers.

One of the most obvious questions for any heuristic is how well it will perform, not only in general, but in any particular instance. This is of particular interest for iterative methods such as those described in the earlier chapters of this book, since if we can determine the quality of a heuristic solution, it will help us to decide when to stop. Given that there is no guarantee of optimality, it is clearly important to have some estimate of how good or bad a heuristic solution is. This is a question that has occupied a fair amount of attention, and it is possible to identify at least three methods of answering it—analytical methods, empirical testing, and statistical inference.

It is not possible within the context of this chapter to give an exhaustive account of these methods, but it is hoped that it will contain sufficient information to be useful to the practitioner. Readers who wish to find out more should consult the various source references that will be given.

7.2 Analytical Methods

7.2.1 Worst-case and average performance analysis

For some heuristics it has proved possible to analyse their operation in such a way as to provide bounds on either their worst-case or their average performance. Thus, for instance, there is a well-known heuristic for the travelling salesman problem due to Christofides which is *guaranteed* to produce a tour no more than 50% longer than the optimal one. (A detailed account of this algorithm can be found in [1, 2].)

As an example of how such an analysis may be carried out, consider the *general knapsack problem* subject to a single constraint:

Example 7.1 (The knapsack problem) *A set of n items is available to be packed into a knapsack with capacity C units. Item i has value v_i and uses up c_i units of capacity. Determine how many of each item should be packed in order to maximize*

$$\sum_I v_i$$

such that

$$\sum_I c_i \leq C$$

Suppose that the items have been sorted into descending order of (v_i/c_i). A 'greedy' heuristic for this problem is to pack as many of item 1 as possible, then fill up what is left with as many of item 2 as possible, and so on. A lower bound on the objective function produced by this method is clearly given by

$$V_H \geq v_1 \lfloor C/c_1 \rfloor$$

where $\lfloor x \rfloor$ denotes the greatest integer which is less than or equal to x.

On the other hand, an obvious upper bound on the objective function is

$$V \leq v_1(C/c_1)$$

so that

$$V_H/V \geq \frac{\lfloor C/c_1 \rfloor}{C/c_1}$$

and writing the denominator as

$$\lfloor C/c_1 \rfloor + (C/c_1 - \lfloor C/c_1 \rfloor)$$

implies that

$$V_H/V \geq 1/2$$

In other words, the worst possible solution to an instance of a general knapsack problem using the greedy heuristic would be half the value of the optimal solution. It is possible to find an instance for which this worst-case actually occurs, but the hope would be that in many cases, the performance would be substantially better than this.

A survey of such techniques can be found in papers by Fisher [3] and Johnson and Papadimitriou [1], but while new analyses continue to be published, the area really lacks a unifying principle and the development has been rather piecemeal. As far as most of the techniques discussed in this book are concerned, there seems to be little or no application of worst-case analysis. In some cases this is not surprising: for instance, local search methods (on which simulated annealing and tabu search are based) have been shown to have no performance guarantee for the TSP, even in exponential search time.

Further, unless these performance bounds can be made fairly tight, such information may be of little help in deciding on the effectiveness of a solution to a particular instance. In view of this, and despite the interesting and often quite subtle mathematics that it entails, worst-case analysis would seem to be of limited practical use.

The analysis of average performance is perhaps even more fragmented; as this entails the assumption of a probability distribution over all possible instances, this is perhaps not surprising. For readers who wish to find out more on this subject, there is a useful survey by Coffman *et al.* [4] on average performance for some sequencing and bin-packing heuristics, while Karp and Steele [5] have reviewed some of the results obtained for the TSP. Again, however, there is little of relevance to the techniques presented in this book, and of course, such analyses are open to the same criticism as worst-case analysis—they still provide little help in assessing the performance of a heuristic in a *particular* instance.

A fortiori, from the point of view of furnishing a *stopping criterion* for an iterative heuristic (such as simulated annealing, tabu search or a genetic algorithm) in a particular instance, such methods are also of little value.

It may be that further developments in these areas will change these rather pessimistic conclusions. Unless or until such developments take place, readers who wish to determine the quality of a solution obtained by one of the techniques described in this book, or who wish to define a stopping criterion for a particular application, will find another approach more useful.

7.2.2 Bounds

One way of trying to decide on the quality of a particular heuristic solution might be to try to obtain a good lower (or, if maximising, upper) bound for the instance in question. Typically this might entail using some form of *relaxation* of the problem to one which can be solved optimally rather more easily. In other cases, bounds can often be developed by fairly straightforward conditioning arguments.

As an example of the first approach, there is the well-known *assignment* relaxation of the travelling salesman problem, which has been exploited in a number of ways over the last three decades. The solution to the assignment problem (AP), as described in chapter 1, can be represented as a permutation $\{\pi_1, \ldots, \pi_n\}$ of the numbers $\{1, \ldots, n\}$, and such a solution can be found by a polynomial algorithm. If an instance of a TSP is solved as if it were an AP, there will typically be sub-tours in this solution. Clearly any valid tour will involve breaking these sub-tours and re-connecting them by links that can be no shorter, so the total distance represented by the AP solution must be a lower bound on the distance of a TSP solution. It is possible to refine this lower bound [6], and to use other relaxations of the TSP (e.g. see [7]). The class of Lagrangean relaxation methods has of course already been covered in detail in chapter 6.

Another general approach to finding bounds is to condition on some partial aspect of a solution. For instance, one of the simplest bounds for the TSP is based on the necessary (but not sufficient) condition that every city must be connected to exactly 2 other cities. Thus a valid lower bound on the optimal tour length is given by

$$\frac{\sum_{i=1}^{n}(d_{i,[1]} + d_{i,[2]})}{2}$$

where $d_{i,[k]}$ is the distance of the kth nearest city to city i.

As another example, consider the problem of finding a valid lower bound for an instance of the $n/m/P/C_{max}$ flowshop sequencing problem. In this problem, n jobs with processing times p_{ij} for job i on machine j have to be sequenced in the same order for each of the m machines, so as to minimize the *makespan* (C_{max}) which is the completion time of the last job on the last machine. Suppose we condition on the first machine. Then it is obvious that

$$C_{max} \geq \sum_i p_{i1}.$$

This can be further improved by noting that whichever job is sequenced last must still be processed on machines $\{2, \ldots, m\}$, so that

$$C_{max} \geq \sum_i p_{i1} + \min_i \sum_{k=2}^{m} p_{ik}.$$

Finally, machine 1 may not be the 'bottleneck', so by conditioning on machines $2, \ldots, m$ in turn, further improvements may be made. Such arguments are frequently used in the context of sequencing and scheduling problems.

In other cases, finding a bound may be a problem-specific affair involving no little ingenuity. But, assuming that a bound can be obtained in some fashion, it is obvious how to use it: the bound forms one end of an interval, the heuristic solution is the other. If this interval is relatively short, we can be confident that the heuristic solution is a good one. Again, if we are interested in a stopping criterion (for a genetic algorithm, say), once the interval between the lower bound and the best-so-far solution has reached a specified threshold, the iterative procedure can be stopped.

7.3 Empirical Testing

It is common practice when reporting a new heuristic to compare its performance with that of existing techniques on a set of *benchmarks*. Sometimes these problem instances will have arisen in a real situation, but real data are scarce, and may represent only a small fraction of the possible population of instances, so benchmarks are often generated by randomly sampling from what is *assumed* to be the population of instances of problems of the appropriate type. For example, Taillard [8] has provided a set of 120 problems of varying sizes for flow-shop, job-shop and open-shop scheduling problems. The processing times for a given job/machine combination were generated by randomly sampling a U(1,100) distribution. There is also a collection of instances of a whole variety of problems available by electronic mail, as described by Beasley [9].

By testing a heuristic across a wide range of problem instances, it is hoped that an idea can be gained of how well it performs in general, and in what circumstances it will do relatively well or relatively badly. To do this properly means using a rigorous experimental design on different levels of the various problem parameters. The analysis of the results can then be carried out using standard statistical methods such as analysis of variance (Anova), or non-parametric equivalents such as the Wilcoxon or Friedman tests. In fact, as the necessary assumptions for a traditional Anova investigation may often be invalid, non-parametric tests are usually preferred. A thorough study of such methods in the context of the TSP has been carried out by Golden and Stewart [10]. Although they deal only with the one problem, what they have to say is of relevance to any other combinatorial optimization problem.

Such approaches are appropriate for finding out what happens 'on the average', since they are tests of location. But it may be of interest to know about the dispersion or other statistical measures, especially if we are comparing two or more heuristics together. For example, if heuristic A has better average (mean or median) performance than heuristic B, but its standard deviation is greater, or its distribution has greater skewness, it is not clear which is the better to use in general. The better average performance of heuristic A may be offset by the fact that there is a greater risk that it will perform badly.

Golden and Assad [11] have suggested a decision-theoretic approach for this situation. Briefly, their argument is that we should explicitly formulate the *utility function* for each heuristic in the form

$$u_t(x) = \alpha - \beta e^{tx}$$

where α, β, t are non-negative parameters. α and β can be chosen arbitrarily, while t provides a measure of risk-aversion.

They fit a *gamma* distribution (which was experimentally justified in all cases they examined) to the data—empirical values of the random variable measuring the deviation of a heuristic solution from a lower bound. This leads to a simple formula for the expected utility of a heuristic as a function of the various parameters, which can then be used to compare the heuristics over a range of risk-values.

A related problem is the decision between heuristics which have different time complexities. White [12] has suggested extending the decision-theoretic approach by incorporating the expected computation time into the utility function. However, despite these interesting suggestions, there is little evidence in published work that researchers are making much use of them.

A further problem with empirical testing is the difficulty of ensuring, even with randomly generated instances, that the benchmarks are sufficiently representative of real problems. In the case of various machine-scheduling problems, for example, there is some evidence [13] that structure of some kind is often present in real problems, and that this structure makes problems easier to solve, even if the method used does not set out to use that structure. The author noticed a similar effect some years ago [14] in investigating real vehicle routing and scheduling problems: for the particular situation under examination, the 'customers' were not randomly scattered but clustered. Furthermore, by randomly generating many problem instances with the same characteristics, it was empirically observed that simple heuristics performed relatively much better, compared to a 3-optimal technique, in the presence of this structure than they did on problems with randomly scattered customers.

A more recent study of the performance of genetic algorithms on flowshop sequencing problems [15] showed a similar effect. Structure was imposed on the problem instances generated, by requiring either that there should be a gradient of job processing times across machines, or that there should be correlation between the processing times of jobs on the same machine. It was again observed empirically that simple techniques, such as neighbourhood search and a greedy type of constructive heuristic, performed almost as well as GAs and simulated annealing for these cases, although they were comprehensively beaten on totally random instances.

A similar effect can be observed in the instances of the knapsack problem used to test the ANN approach in chapter 5 (see Table 5.2). In the case of random instances, there was little to choose between the simple (but fast) greedy heuristic and the more sophisticated (but slower) approaches in terms of solution quality; the case was quite different for more homogeneous instances. Again, in the extended series of experiments on simulated annealing performed by Johnson *et al.* which were discussed in chapter 2, attention was also drawn to the possibility that algorithm performance could vary according to differences in particular features of problem instances.

Such results suggest that the choice of a heuristic for a given problem instance is not clear-cut. Some of the factors that should enter into such a decision are the problem size, the problem structure, and the cost of using a sub-optimal solution rather than an optimal one, as well as any data relating to the performance and computational cost on similar problems of the heuristics under consideration.

7.4 Statistical Inference

The problem of estimating a parameter of a statistical population from sample information is a familiar one to statisticians, and it is perhaps surprising that it was so long before statistical techniques of estimation were applied in combinatorial optimization. It appears that the first attempt to use statistical inference was by McRoberts [16]. His work has been extended and developed by Dannenbring [17], and in several papers by Golden and colleagues [18, 19, 20].

The essential results come from the statistical theory of extreme values and were first proved over 60 years ago by Fisher and Tippett [21]. Consider the situation where we take n independent samples, each of size m, from a population whose minimum value a we wish to estimate. Suppose the smallest value in sample i is v_i; then it can be shown (see, for instance, Gumbel [22] for the details) that as $m \to \infty$ the distribution of v_i approaches the Weibull distribution

$$F(x) = P[X \leq x] = 1 - \exp\left\{-\left(\frac{x-a}{b}\right)^c\right\}$$

where a, b, c are non-negative parameters, and $x \geq a$.

Strictly speaking this result applies to continuous distributions, whereas the distribution of the objective function values for a combinatorial problem is clearly discrete. However, the size of the population for most combinatorial problems is such that the distribution is 'approximately' continuous, and empirical evidence is strong that the results can be applied with a high degree of confidence.

The parameter a is clearly the one of greatest interest, as it represents the minimum value realisable by the random variable x. At first, although is is clear how this could be applied to a situation where we generate a large number of random solutions to a combinatorial problem, it is not so obvious how this applies to evaluating a heuristic solution. Golden and Alt [20] argue that each time we apply a heuristic, we *implicitly* sample a large number of possible solutions, of which we find the minimum v_i. They report on some extensive computational experiments for the case of the TSP which showed empirically that this argument appears to be valid. They also found that if different solutions were obtained by applying the heuristic several times there was no reason to reject a hypothesis that these solutions were independent.

Given n independent solutions obtained in this way, it is possible not only to find a point estimate for the overall minimum, but also to find a confidence interval. Golden and Alt argue in the following way: from the Weibull formula, it is clear that

$$F(a+b) = 1 - e^{-1}$$

so that if

$$w = \min_i v_i$$

$$P[w \geq a+b] = \{1 - F(a+b)\}^n = e^{-n}$$

and hence

$$P[w \leq a+b] = 1 - e^{-n}.$$

Since $w \geq a$ (by definition), this can be re-arranged in order to give a $100(1 - e^{-n})\%$ confidence interval:

$$w - b \leq a \leq w.$$

There remains, of course, the problem of estimating a and b. In [20] it is suggested that all three Weibull parameters should be estimated by the method of maximum likelihood.

A more 'natural' confidence interval has been proposed by Los and Lardinois [23]. They observe that for any scalar T,

$$F(a + b/T) = 1 - \exp\{-T^{-c}\}$$

From this, using the same argument as Golden and Alt, it is possible to deduce a $100(1-\alpha)\%$ confidence interval:

$$w - b/T \leq a \leq w$$

where

$$T = \{-n/\ln\alpha\}^{1/c}.$$

Empirical evidence for the TSP and some other problems gives no reason to doubt the validity of this approach, although there is clearly scope for further research. However, despite the apparent simplicity of the approach, and its promising performance, there are few reported applications in the literature.

7.5 Conclusion

This book has introduced the reader to a number of very effective and interesting heuristic techniques. Some of them are clearly still in their infancy, and there is a lot of work still to be done. Without having set out explicitly to do so, we may well have suggested many research topics ourselves! We have also stressed the importance of evaluating heuristic performance, and given some suggestions and guidelines that should be followed. At the very least, we hope that this work will help both students and practitioners to solve their own problems more effectively.

References

[1] D.S.Johnson and C.H.Papadimitriou (1985) Performance guarantees for heuristics. *In* [2], 145-180.

[2] E.Lawler, J.K.Lenstra, A.H.G.Rinnooy Kan and D.B.Shmoys (Eds.) (1985) *The Traveling Salesman: A Guided Tour of Combinatorial Optimization.* John Wiley and Sons, Chichester.

[3] M.L.Fisher (1980) Worst-case analysis of heuristic algorithms. *Man.Sci.*, **26**, 1-18.

[4] E.G.Coffman, M.R.Garey and D.S.Johnson (1984) Approximation algorithms for bin-packing: an updated survey. *In* G.Ausiello, M.Lucerstini and P.Serafini (Eds.) *Algorithm Design for Computer System Design.* Springer Verlag, New York.

[5] R.M.Karp and J.M.Steele (1985) Probabilistic analysis of heuristics. *In* [2], 181-205.

[6] N.Christofides (1972) Bounds for the travelling salesman problem. *Ops.Res.*, **20**, 1044-1056.

[7] M.Held and R.M.Karp (1970) The traveling salesman problem and minimum spanning trees. *Ops.Res.*, **18**, 1138-1162.

[8] E.Taillard (1992) Benchmarks for basic scheduling problems. *EJOR* (to appear).

[9] J.Beasley (1990) OR Library: distributing test problems by electronic mail. *JORS*, **41**, 1069-1072.

[10] B.L.Golden and W.Stewart (1985) The empirical analysis of heuristics. *In* [2], 207-249.

[11] B.L.Golden and A.A.Assad (1984) A decision-theoretic framework for comparing heuristics. *EJOR*, **18**, 167-171.

[12] D.J.White (1990) Heuristic programming. *IMAJMABI*, **2**, 173-188.

[13] A.D.Amar and J.N.D.Gupta (1986) Simulated versus real life data in testing the efficiency of scheduling algorithms. *IIE Transactions*, **18**, 16-25.

[14] C.R.Reeves (1975) *Algorithms for vehicle scheduling in an industrial context.* M.Phil dissertation, Lanchester Polytechnic.

[15] C.R.Reeves (1992) A genetic algorithm for flowshop sequencing. *Computers & Opns. Res.*, (in review).

[16] K.McRoberts (1971) A search model for evaluating combinatorially explosive problems. *Ops.Res.*, **19**, 1331-1349.

[17] D.Dannenbring (1977) Estimating optimal solutions for large combinatorial problems. *Man.Sci.*, **23**, 1273-1283.

[18] B.L.Golden (1977) A statistical approach to the TSP. *Networks*, **7**, 209-225.

[19] B.L.Golden (1978) Point estimation of a global optimum for large combinatorial problems. *Comms. in Stats.*, **B7**, 361-367.

[20] B.L.Golden and F.B.Alt (1979) Interval extimation of a global optimum for large combinatorial problems. *NRLQ*, **26**, 69-77.

[21] R.Fisher and L.Tippett (1928) Limiting forms of the frequency distribution of the largest or smallest member of a sample. *Proc.Camb.Phil.Soc.*, **24**, 180-190.

[22] E.Gumbel (1958) *Statistics of Extremes.* Columbia University Press, New York.

[23] M.Los and C.Lardinois (1982) Combinatorial programming, statistical optimization and the optimal transportation network problem. *Trans.Res.B*, **16**, 89-124.

Index